PRINCE EUGENE'S WAR IN HUNGARY 1716-1718

By
Friedrich Wilhelm Karl von Schmettau

Translated by
George F. Nafziger

Original work:
VIENNA
Chez Gräffer le Jeune
·············
1788

Prince Eugene's War in Hungary 1716 - 1718 by Friedrich Wilhelm Karl von Schmettau
Translated by G. F. Nafziger
Cover: Prince Eugene of Savoy by Jacob van Schuppen
This edition published in 2023

The Nafziger Collection is an imprint of

Winged Hussar Publishing, LLC
1525 Hulse Rd, Unit 1
Point Pleasant, NJ 08742

Copyright © Winged Hussar Publishing
ISBN 978-1-950423-93-4 Hardcover
ISBN 978-1-958872-12-3 Ebook
LCN 2023931093

Bibliographical References and Index
1. History. 2. 18th Century. 3. Central Europe

Winged Hussar Publishing, LLC All rights reserved
For more information
visit us at www.whpsupplyroom.com

Twitter: WingHusPubLLC
Facebook: Winged Hussar Publishing LLC

Table of Contents

OPINION OF THE EDITOR

This work which we give to the public must naturally excite its interest; it acquires a new price by circumstances. All that which is relative to the preceding discussions between the House of Austria and the Ottoman Porte must be greeted favorably since it is only there that one can find the reasons for the war that occupies today the arms of the two powers.

We believe we should not stop at the moment of the Peace of Passarowitz. This treaty, written in Latin, is necessary to understand the reasons for the following war and we give it here in translation.

The author of this Hungarian war has not judged it appropriate to make himself known, but the purity of his style, his manner of relating his story, and the exactitude of the facts can leave no doubt of his talents and of the sources from which he has drawn.

Frederick Wilhelm Karl von Schmettau

Prince Eugene of Savoy by Jacob von Schuppen

HISTORY OF THE 1716 CAMPAIGN

It had been some time since the Ottoman Porte has sought an occasion to make war against the Venetians, counting no doubt the regaining from the Porte that which it had lost in the war they had had against the Emperor. The conquest of the Peloponnese appeared to it to be very easy and very proper at the same time as compensation for the fortresses and the lands that it had been obliged to cede in Hungary by the Treaty of Karlowitz. It began arming on sea and on land in a formidable manner. The Imperial Court was initially very disturbed concerning the destination of this storm, not knowing where it would fall. On the other side, the return of the King of Sweden to Pomerania, the preparations of the Muscovites and Danes, of the Saxons and Prussians, produced a fear that would re-ignite in the heart of the Empire, and that His Imperial Majesty could not receive the assistance from the states of the Empire, if a war was being fought at their doors, and in the case that the Turks should attack in Hungary. In addition, there was the concern that this monarch should find himself obliged to divide his own forces to put himself in a posture inside the Empire and to repulse an external enemy. However, all of this tumult, which the princes aligned against Charles XIII in the north, had not attracted the attention of the Imperial Court for a long time. It was left to the armed kings the need to pursue their quarrel, making on their side the necessary dispositions to resist the Turks.

The Ottoman Porte was not unaware of the offense that these extraordinary armaments caused the Emperor and undertook to disabuse, or to amuse him with pretty protestations; so that neglecting to arm himself he could not send relief to the Ottomans, nor delay the progress of Ottoman arms. To this end the Porte sent an Aga to Vienna, with instructions to persuade the Emperor that the Grand Seigneur was only taking up arms so as to avenge himself against the Venetians, who he claimed, had made a thousand snubs to Turkish ships, in capturing many which they had sold and made slaves of the owners. The Aga left, at the end of February, from Constantinople to go to Vienna. The Emperor sent before him the Commissioner General to escort him from the borders of the Turkish Empire to the Austrian Capital. The Court Interpreter, Schmidt, was also sent to Aga Ibrahim, who arrived at Vienna at the beginning of May, accompanied by about 20 people of his suite, a guard of 20 Imperial soldiers, and an interpreter from Belgrade, with that of the Imperial Court, and the Commissioner. All of their baggage was carried in 12 wagons.

On 13 May the Aga was led to the audience of Prince Eugene with ceremonies similar to those observed on a similar occasion in the time of the Emperor Joseph, though on this occasion there was a bit more magnificence. The Turkish envoy forgot nothing to prevent the Viennese Court from taking part in the quarrel that the Porte had with the Venetians. He submitted a letter from the Grand Vizier to Prince Eugene, which contained protestations of friendship on the part of the Grand Seigneur towards His Imperial Majesty. The Emperor offered, by Sr. Fleischmann, his resident in Constantinople, his mediation to His Highness and Prince Eugene also proposed it to the Grand Vizier in the response that he made to his letter. As the Divan wished war, it could make no positive response to the offers of mediation, and it continued to arm.

As a result, the Emperor felt himself obliged to take another tone with the Turks. He had Sr. Fleischmann say to the Grand Vizier that if the Porte made war against the Venetians, His Imperial Majesty could not exempt himself, as a guarantor of the Treaty of Karlowitz, from declaring war on him. Prince Eugene also said the same to Aga Ibrahim, who was leaving after having performed his embassy. On the threats made by the Viennese Court, the Grand Seigneur sent troops into Hungary and ordered that they work to put its fortresses in a state of defense.

However, the Turkish Fleet, having set sail from the Dardanelles landed on the little Island of Tine an army of 50,000 to 60,000 men, which penetrated into the Peloponnese by the Isthmus of Corinth, besieged and took Napoli de Romanie.[1]

---

[1] Translator: Identifying 18th century Greek and Balkan cities whose names have been presented in this work in French was not attempted except for those obvious names. In addition, several regional forces, i.e. Recines, are referenced and they too could not be

Venice directed all of its efforts to put itself into a state of defense. This republic did not maintain in peacetime more than 6,000 men, both infantry and cavalry. When it wished to augment this little army, it had great difficulty in finding recruits; the greater part of the men being persuaded that they had been discharged at the end of the war and the Republic had the reputation of employing on its galleys, as galley slaves, the soldiers that it no longer needed ashore. This idea, which the Venetian officers could not destroy, some ruses which they used augmented the natural difficulties of finding men who wished to fight in an unknown land, and in the conservation of which they had little interest. The Venetian Senate could raise some regiments in Switzerland and in the Grisons. But as that would not suffice to stop the Turks, the Republic had recourse to the German princes, who for money were always ready to sell soldiers. In this manner Venice quickly formed an army of 30,000 men. Its fleet was reinforced by the galleys of the Pope, those of Malta, and several French ships. Nothing major passed between the Turks and the Venetians during the rest of this campaign, which was nothing but a prelude to others.

The Emperor seeing well that war with the Turks was inevitable, worked to finish the difficulties that he encountered in the affair in the Lowlands, which were still in the hands of the English and Dutch, who ruled them in common. These two powers did not refuse to return these lands to the Emperor, but assured him perpetual possession of them, it was a question of guaranteeing Holland from all threats; and for that it required that things be arranged so that these lands always serve as a barrier between France and Holland. The death of Louis XIV, which occurred at the time of this affair, eliminated most of the difficulties. A treaty was signed on 15 November. It contained 29 articles in which the Dutch appeared to have exhausted their concerns.

"It is stipulated here that the provinces and cities that formed otherwise the Spanish Lowlands, shall remain united to the domain of the House of Austria in Germany, and to its successors in perpetuity such that they can never be ceded or transferred, under whatever title may, to the Crown, nor any

identified in their modern English form.

Prince or Princess of France. That the Emperor and Holland shall maintain at their expense, always, a corps of 15,000 to 30,000 men; that this number, in the fear of war, shall be augmented to 40,000 and beyond, for which the Emperor shall pay three fifths. That the garrisons of Namur, Tournai, Menin, Furnes, Warneton, Ypres, and the Kenocque Fort (Fort Knoffe), shall be formed only of Dutch troops; that those of Dendermonde shall be common, but that the governor of that city shall be named by the Emperor alone, on the condition that he shall swear an oath to the States General, and to the other officers, such that nothing he might do be prejudicial. That the staffs of the fortresses where there is only a Dutch garrison, shall be at the nomination and at the appointments of the Republic, on the condition that they not be disagreeable or suspicious to the Emperor, to whom they shall swear an oath to guard these fortresses faithfully for the sovereignty of the House of Austria. That with regards to religion, all things shall remain in the Austrian Lowlands on the footing as they were during the reign of Charles II, and as to the Dutch troops, they can exercise everywhere that they may find themselves in garrison, their religion; but in the particular locations and in proportion to the number, which they are assigned by the magistrate of each city, without having the right to place exterior marks of their faith on their churches. That the convoys of munitions of war and provisions from Holland, when there is an appearance of a rupture with the neighbors, shall not be subject to the rights of tolls belonging to the Emperor; but may only be visited for preventing frauds and abuses. That the actions of the Republic that may be done to change garrisons shall be organized and regulated with the General Governor of the Lowlands, as well as the reinforcement of the troops which one might believe necessary to place in the fortresses that find themselves in danger of being attacked or surprised; as also for that which regards the repair or construction of fortifications whose costs shall not be charged to the Emperor, without having his consent. In the case of enemy armies entering the Brabant, the Republic shall post its troops on the Demer from the Scheldt to the Meuse, constructing necessary lines and inundations, all in agreement with the Governor General; and that for this end the Emperor shall cede to Holland as much territory of Austrian Flanders as may be necessary for these inundations from the Scheldt to the sea. That the limits of the States General in Flanders, beginning at the sea between Blanckenberg and Heyst; that from this latter place they shall continue on the Drillhoëck from Swarlessuys, from there to Fort Saint Donas, which the Emperor cedes in all sovereignty to the Republic, on the condition that the gates of the sluices of this fort shall be removed in time of peace. That from St. Donas the new limits of Holland shall extend to Fort St. Job, to regain their old [lines?], near Middelburg, and then along the Zuidlingsdyck, such as they had been before. That in regard to the city of Pas de Ghent, its limits shall extend 2,000 geometric paces; and for the conservation of the Sas-Scheldt, as well as to facilitate communication between the Brabant and Dutch Flanders, the Republic shall retain all ownership of the villages and dependencies of Doël, of Saint-Anne, and Ketenisse; such that it shall be the same in full sovereignty in the upper quarter of Guelders, the city of Venloo, the Forts of St. Michel and Steffenwaërt, with the ground necessary to augment the fortifications beyond the Meuse, and the Ammanie of Montfort with all of the attached rights; on the condition that the ecclesiastic privileges of the civilians shall be conserved, such as they were under the reign of Charles II. So as to contribute to the costs of the Republic for the conservation of the Austrian Lowlands, the Emperor shall receive every year the sum of 500,000 Dutch écus, which he declares hypothetically on all of the revenues of his new pension; and that to assure the fate of the people who have been governed by the Ministers of Great Britain and the States General, as representatives of the legitimate sovereign, the Emperor confirms and ratifies all of the judgments rendered during their administration on the affairs of the state, justice, police, and finance."

This treaty having reassured the Dutch on the side of France and given the Emperor sovereignty of the Lowlands, this monarch always full of gratitude for the services of Prince Eugene, gave the General Government to His Serene Highness. The patents were expedited on 21 June. This task had always been very considerable and the most desired in the Spanish Court. The kings of this name were accustomed to only giving it to their most favored subjects, for whom they thought they could give no greater expression of affection than in naming them Governor General of the Lowlands. It was, without doubt, with this in mind that Charles VI turned his eyes on Prince Eugene to raise him to this honor. The Marquis de Prié, an Italian, was named his lieutenant to command in his absence. The Government of the Milanese, which Eugene had at the time, was given to Prince von Löwenstein-Wertheim, the same who had been the principal commissioner to the Diet of Ratisbon.

Before acting against the Turks, the Emperor consulted his council many times, so as to do nothing too precipitously. They thoughtfully examined if they should simply declare war or continue down the path of negotiations.

Prince Eugene, avid for new laurels, was among those who most avidly called for war. He represented to His Imperial Majesty: "that he could not fail to declare against the Turks in favor of the Venetian Republic, which had concluded the Holy League with the Emperor Leopold of glorious memory, had fulfilled all of its conditions, in declaring war on the Turks when they made war on His Imperial Majesty. That the glory of the House of Austria was interested here, since it would do wrong, to see such faithful allies be oppressed without assisting them. That in addition, the honor, the interests [of the Empire] were involved, the Hereditary Lands of His Imperial Majesty could not but be exposed by the progress of the Turks on one side or the other. That if they came, by example, to capture the Island of Corfu, which has always been regarded as the pathway to Italy, nothing would prevent them from conquering the Kingdom of Naples, then penetrating into Milanese territory, then into the Tyrol, and attacking the Empire from the west, while their army in Hungary attacked it from the east. That nothing could be expected from the path of negotiation; that the intention of the Porte was not to remain peaceful; but only to amuse the Emperor until it had overwhelmed the Venetians, which it only wanted because it thought they were not in a state to resist them. That there was no reason to fear that the Emperor would be troubled by any Christian princes while he was engaged in such a holy and just war. That in addition, reasons of honor and religion engaged them to remain at peace, that there were others that resulted from the particular constitution where they found themselves. That France, for example, was more interested in recovering from its losses than attempting to take any advantages. That the Duke d'Orléans, who governed this Kingdom, was more occupied with maintaining himself in the Regency, and to guaranteeing himself from its internal factions, than in acting outside [the Kingdom]. And finally, the Court of Madrid has enough to do with the Catalans and that King Philippe is little assured on his throne to think about attacking anyone."

These reasons and many others, no less important, determined the Emperor to declare war. He named Prince Eugene as the General-in-Chief of his armies in Hungary. Large numbers of recruits were sent into the kingdom, new cannons were cast, munitions of war and all sorts of provisions [were accumulated]. The soldiers and officers who had greeted the Peace of Rastadt were recalled and all of the regiments destined to serve in Hungary received orders to hold themselves ready to march in April. Prince Eugene worked incessantly on his crews. It was ordered that there be three army corps in Hungary, the first with 70,000 men, was to be commanded by Prince Eugene in person; the other, with 30,000 men, who were to be commanded by Count Gui von Stahremberg, and the third, of 25,000 by General Heister.

These dispositions were made by a great council of war which was held on 15 March, concerning the first operations that were to be undertaken. Count Gui von Stahremberg, who had left Gratz in Styria to attend, spoke of the preparations by the Turks in a manner to wake up the attention of all of the Empire. He stated that there had been a gathering of infidel troops on the frontiers of Hungary and Transylvania, and of the severe defenses that they had made in bringing in cattle and horses.

The letters from Constantinople indicated that there had been a Grand Divan held, and when it ended most of the Pashas who had attended it had returned to their governments in order to assemble the troops that they were to furnish. These letters added that the Horse Tails had been exposed for several days; that the enrollment of soldiers had been undertaken in the capital with an extreme diligence, and that one worked there with no less effort in making new tents for the troops that would soon be camping in the field.

The Viennese Court redoubled its efforts upon receiving this news. The galleys that they had constructed were equipped and they sent orders into Lower Germany to gather sailors to serve on these galleys, which were destined to act on the Danube. They brought from Nuremberg and Frankfurt more than 3,000 bakers to bake bread for the army. Orders were sent into Hungary to construct bridges over the rivers and to repair the roads, in order to facilitate the movement of troops. In Vienna, many rowboats were constructed and equipped with cannon to serve on the Danube.

From 15 April, the Imperial infantry regiments marched to form the various camps in Hungary; but as the winter was very long this year, and that the grass had not yet begun growing, the departure of the cavalry was delayed until the end of the month.

A few days after Prince Eugene held a review of the 10 companies of the old Lorraine Regiment, which left for Hungary with five other companies of the Bagni Regiment, and a quantity of recruits.

The Turks well-anticipated that Temesvar was the first place that would offer itself to the arms of the Emperor at the first attack, and worked with much diligence to put it into a state of defense. They did not content themselves with only repairing the ancient fortifications, but also constructed new ones. They employed 1,300 Wallachians that they had torn from their land, because they had not been able to pay the enormous taxes imposed on all of Wallachia.

All of the precautions of the infidels augmented those of the Emperor. This monarch seeing that he no longer had need to hold troops in Italy beyond those necessary to garrison the fortresses, recalled the rest. The Wetzel Regiment left for service in Hungary. The Emperor personally reviewed three cavalry regiments: Baden Durlach, Harrach, and Prince of Lorraine (the younger). This review was held on the Leopolidin Island, where the troops were then transported by boat down the Danube. Count von Kaunitz was sent to the Imperial Princes to solicit troops and money. His solicitations were not without result. From 27 June, the three Colleges of the Germanic Corps issued the unanimous resolution that they would support the Emperor with all of their forces. The Pope accorded him an indulgence to tax the revenues of the clergy and extraordinary rights on all of the ecclesiastic goods of his Hereditary States. An Apostolic Nonce had them posted on all of the churches in Vienna.

These dispositions of the Empire and of the Court of Rome were a good augur for the Imperial Court and caused great joy there, which was diminished a bit by the loss of Count von Guttenstein, Governor of Prague, a worthy and experienced officer, and who had by his wisdom won the affection of the Bohemians, and who had been of great use in this land for the interests of the Emperor. The Emperor publicly praised this general, as did Prince Eugene, who had always held him in great esteem. Baron von Pickingen was named to replace him at Prague.

In the meantime, two things occurred that raised, a bit, the ardor of the Turks arguing for war. The first was two fires arriving one after the other at Temesvar, where about 40 houses were consumed, and another at Belgrade, where 30 barges loaded with grain and other provisions were burned. The Turks, naturally suspicious, saw this as a sinister portent and claimed to be intimidated, when Baron von Löffelholz, Governor of Peterwardein, had advanced with the troops that were under his orders, as far as Mitrowitza, on the Save, beyond the limits that separated the two empires, and the Turks had not opposed him. Mitrowitza was, all the same, an important post. It is situated on a piece of ground that is about five leagues long and as many wide; and the fortress is at the confluence of the Save and the Danube, no farther from Belgrade than the width of the Save. A large ditch had been dug between the two rivers to serve as the limits between the two Empires; and this ground, in which there are three towns or villages, the principal of which is Mortrowitza (the others are Barils and Semlin) belongs to the Turks.

The Pasha sent his complaints to Baron von Löffelholz for this contravention, as if the Sultan, his master, had not already made various violations of Treaty of Karlowitz, and as if he had not been the first who had violated the peace established by this treaty.

Löffelholz responded that he was not posted in this region by the order of the Emperor; that he would only remain there until the return of the courier that the Emperor had sent to his Resident in Constantinople, to have a positive response on the part of the Ottoman Porte; and that if the courier carried a favorable response, he would immediately retire. He added that he also had complaints to make on his side for various infractions of the Treaty of Karlowitz that the Turks had made during the previous month, among others which they had committed recently, when their ships had fired on Imperial troops moving down the Save, and had fired on them continually with their cannon, perriers, and muskets, even though these troops had done them no ill, since they were in march to the Emperor's territory.

The Pasha, who had not expected such a parallel recrimination, sought out those who had manned the boats that had fired on the Imperials, and had them impaled. However, the Imperials did not regard this execution as anything more than the result of his spite, and they knew well how his court looked upon this. In effect, one had certain word that the Turks were continuing more and more their preparations. It was known that a corps of Ottoman troops was forming near Belgrade; that the Turks had reinforced the garrison of this city such that it came to about 40,000 men, and that 60,000 Spahis were camped near the frontiers of Transylvania, where they were to be joined by 30,000 Janissaries.

At the end of May the Württemberg Regiment arrived at Vienna by water, coming from Swabia. Prince Eugene, accompanied by several generals, reviewed it and found it one of the best of the Imperial Army. Two days later Captain Schwendimann arrived from Hamburg with 240 sailors destined to serve in the armed frigates that had been constructed to act on the Danube, which he was to command. The next day the Imperials launched the fifth of these ships, armed with 48 cannon. Prince Eugene was present as it was launched. At the beginning of June, Eugene's baggage left. But he did not go with them, as his presence was still necessary in Vienna.

The Emperor had signed an offensive and defensive alliance with the Venetian Republic on 13 June. This time is remarkable not only for the alliance formed with the Venetians, but for the birth of an archduke which was born to the reigning

Empress. This young prince was the object of the good wishes of not only Their Imperial Majesties, but of all the Empire, and almost all of Europe. He was baptized and named Leopold. The Emperor conferred on him the Order of the Golden Fleece the day after his birth. Sadly, the child died shortly later and the Germanic Corps was deprived of the hopes it had conceived from this happy birth. This loss was all the greater as there were no male heirs in the Imperial family.

However, the anger of the Turks, informed of the measures being taken by the Emperor to make war on them, came to a boil. They locked up the Emperor's Resident to Constantinople and the courier that the Viennese Court had sent to learn the last resolutions of the Porte. They restrained themselves, nonetheless, and contented themselves with advancing their grand army into Hungary under the orders of the First Vizier. It contained about 120,000 men. The general pretended to march on Dalmatia, but suddenly he shuffled towards the Save and Belgrade, where me made a detachment of 3,000 men to cover Temesvar, where he saw, by the position of the Imperial troops, that a storm was about to fall. He sent over the Save a corps, barely less considerable than the preceding, and had it camp on some ground that belonged to the Sultan. A few days later, he moved there with the entire army and threatened severe punishment to his troops that might commit any act of hostility before the Christians had started a war. He imagined that he could thus persuade that it was the Imperials, who were those who broke the Treaty of Karlowitz and not the Muslims.

Finally, Prince Eugene, whose departure had frequently been deferred, left Vienna on 1 July. His regiment of dragoons, which had been called from the Lowlands, followed him into Hungary. He slept on the 3rd, at Buda, where several generals attended him to receive orders, which their having been given, they left early to return to the regions where their duty called them. The next day Prince Eugene arrived at Walcowar, where he spent the night; and having continued his way, he arrived, on the 9th, at Futack, where he held a review of the troops under the orders of Count von Palfy and went from there to the general camp near Bechz, where all the troops had orders to go. The rumor spread that the Turks were preparing to cross the Save, so Prince Eugene commanded Baron Langlet, with 500 men, to dispute their passage, but they did not dare to attempt a passage. The detachment of Mr. Langlet having grown to 3,000 men, by reinforcements sent to him by Prince Eugene had taken care to send him, the commandant was warned to form an enterprise on Rathza, an important post on the Save in the region where the Drina entered that river. A corps of 6,000 Turks appeared to wish to oppose Langlet, but he continued pursuing his point and captured a post under the nose of these 6,000 men, who made no effort to block him. They recrossed the Drina when they saw him and abandoned to him four large boats loaded with beams and planks; and the men who manned those boats quickly fled and recrossed to the other side. Colonel Langlet raised a fort in this region, to cover the troops that he put in Rathza.

While the Imperials deployed to act against the infidels, the Emperor had done in Vienna the things worthy of his piety to draw the benedictions of Heaven on his arms. There was, on 11 July, a general procession of all of the secular and regular clergy, of the magistrates, and the significant members of the Court and the city of Vienna. The procession left the Church of the Shoeless Augustinians (Augustinekirche), and after a great parade they returned to the Cathedral of St. Stephen (Stephansdom), where Count von Kolonitsch, the Bishop of Venice, celebrated a mass, where the Emperor and Empress assisted with great marks of piety.

The day after the ceremony another no less devoted ceremony was held. This was the benediction of the seven warships that had been constructed on the Danube. The Archduchess assisted in this ceremony and were saluted by three salvos of the ships' artillery, which were named: *St. Maria, St. Leopold, St. Johann, St. Karl Boromee, St. Elizabeth, St. Stephan,* and *St. Friedrich.*

Two days later three of these ships set sail for Hungary. It was necessary to remove two arches of the bridge over the Danube to allow them to pass. On their departure, they fired a general salvo of all of their artillery.

In the meantime, a sort of manifesto was distributed by the Grand Seigneur, in the form of a letter addressed to all of the

Pashas of his empire, where he exposed the reasons in which he was engaging in a war with the Emperor. He complained that the Emperor had been the first to violate the Treaty of Karlowitz, in declaring for the Republic of Venice. He said in this letter, "that Ibrahim Aga had been dispatched to Vienna to certify with this court, that all of the preparations that they made had no object but to attack the Venetians; that it was true that Sr. Fleischmann had then offered the mediation of his master, that the President of the Council of War (Prince Eugene) had also made this offer, as had the Ambassadors of England and Holland, but having been accepted, the Venetians had continued their hostilities; that their fleet had wintered at Corfu, a city that was only two miles from the shore and frontiers of the Ottoman Empire; that finally, on the notifications received from various locations and all of the confines of Hungary extraordinary war preparations being made by the Emperor, such as levies, assemblies of troops, establishment of magazines, construction of ships on the Danube, etc.; one had once again pressed his Resident to declare himself; that one had given him a delay of 30 days to make a precise response, and at the end of two months of silence he contented himself to say verbally, that the Ministers of the Porte had not positively responded to the Emperor's offer of mediation, nor on the letters submitted by Prince Eugene to Aga Ibrahim; that he had added several equally feeble reasons, the conclusion of which had been that the Emperor had ancient engagements with the Venetians and that he was obliged to support their interests. That finally, he had said that as since he had been given no precise response, a longer stay in the Porte was useless, and he had requested permission to return to the Court of the Emperor, his master." This manifesto ended with a copy of the letter that Prince Eugene had written to the Grand Vizier and by an order to all of the Pashas and other governors of the Ottoman Empire, to hold themselves in a state of defense, without the least violation of the Treaty of Karlowitz; and to this end, the Sultan declared that although his troops were moving on Belgrade and his ships were sailing on the Danube, his plan was not to attack the Emperor; but only to defend his territories and his subjects of the Ottoman Empire. One found in this manifesto a certain art and a turn of expression that caused one to suspect that the author had not always lived with the Turks.

One surprised, from time-to-time, Turkish spies in the Imperial camps and one was captured in that of the Count von Palsy declared that three others, disguised in rags, had been sent to discover the Imperial camps and forces. His sincerity did not save him and he was still impaled. One also surprised an Hungarian officer, who come from Constantinople and had arrived at Temesvar seeking to cross the Marosch. Others were found who had commissions from the Porte to assemble the German, Polish, Hungarian, and Swedish deserters who remained in the region, who were to be commanded by a foreign general, whose name was never discovered, but which little effort had been made to discover.

On 27 July, Prince Eugene moved to Peterwardein, where he reviewed seven infantry regiments and the artillery, which was under the command of the Count de la Croix. That day several deserting Turks were questioned by the order of His Highness. One learned from them that the Aga of the Janissaries had arrived in Belgrade; that he had been followed by Dgebi-Pasha, and that the troops that were under his orders were formed in a camp which had been marked out for them below Belgrade, and was about a league and a half in length; and that two days later the Grand Vizier had arrived with 30,000 Tartars, from whom a few detachments had been made and sent forward. They added that this army had orders to rest six or seven days, and then to move into their old entrenchments at Semlin, and to make raids from there below Sirmich and as far as was possible, so as to pillage the land according to the Turkish practice. However, they said that these troops had the order to not touch the fields below, seeking to spare them and use them for forage, which they needed greatly. It was not the same with regard to the various Imperial camps. Everything was in abundance there, such as food and forage, nothing was lacking since one had set free the serfs from the feudal rights and commitments if they brought in provisions.

The Turks abandoned the bridge that they had begun building over the Danube to construct another on the Save, which they completed on the 22nd. They employed 3,000 workers, supported by 1,000 Janissaries. A force of 70 ships, as many saiques[2] as brigantines or frigates, had arrived at Belgrade, to which the peasants of the fields around Sirmich had brought a thousand wagons of forage. The generals of the Ottoman army published in their camps very strict orders against anyone going beyond the barriers and against any guard who might allow them to do so. This order was rigorously enforced, such that six Rasciens having crossed the bridge near Arrad, were immediately shot by the advanced guards of this army.

About this time the Imperial field artillery arrived at Prince Eugene's camp and the Imperials worked on constructing two bridges near Peterwardein to allow the passage of what was necessary to support the army over the Danube, because Prince Eugene had orders to draw the Grand Vizier from his lines and to give battle to him. The Prince had no difficulty in succeeding in this regard. The Turkish general was equally desirous of battle and was ready to come halfway.

The Grand Vizier was named Hali. He was a man of fortune, who did not lack either spirit or heart, but he lacked experience, no more than the troops he commanded. The reigning Sultan, Achmet III, had given him his daughter in marriage, though she was at the time only eight years old. He liked this First Minister, and it was at his persuasion that he had broken with the Venetians, and refused any accommodation. This Hali was a great enemy of Christians and capable of executing great evil towards them. When Count von Breuner was brought to him, Breuner having been taken prisoner, he told him that he absolutely wished to cut off his head. However, he turned from this because he had been promised 100,000 florins, and by the pressing prayers of Maurocordato, the interpreter from the Porte, and subsequently Hospodar of Wallaching. In addition, Hali perfectly understood the intrigues of the Seraglio, but for the command of the armies, it was an art for which he had no taste. His boldness to anticipate Eugene in battle was only the result of his ferocity and the hatred he had for Christians. When one represented to him that an action against a general who did not know how to lose a battle, and who joined to it a vast genius, and great experience, required great

---

[2] Translator: As there is no entry in *Petit Robert*, the French equivalent of *Webster's Dictionary*, this is apparently a Turkish word.

circumspection, he responded arrogantly, that though the Grand Vizier[3] of the Christians was a good general, the Grand Vizier of the Muslims could become a better general than him at the expense of his rival. With these ideas, he attempted to dissipate the impressions of fear that the name of Prince Eugene produced in the spirit of his troops. There was not a Turkish soldier who did not know who had crossed at Zenta. Several of them were there and they knew the details of his campaigns in Italy, Bavaria, and Flanders too well to challenge the success of the battles that they would begin with him.

However, the Grand Vizier stood by his guns. He wished to fight. Already half of his army had crossed the Save and the rest followed closely. Prince Eugene, informed of this, wished to examine the Turks and know, more or less, the number that had crossed the river. Count Johann von Palfy forcefully requested from Prince Eugene the honor of commanding the detachment that would be sent to reconnoiter the Turks, which would consist of 1,600 German, Hungarian, or Rascein horse. Eugene gave him the honor. Palfy left and having crossed the Danube came very close to Karlowitz, where he discovered a corps of Turkish cavalry. He could not come close enough to accurately estimate their number, but he could well judge that they were very superior to his own. Based on this he sent word to Eugene requesting reinforcements. Eugene sent two cavalry regiments, to which he joined the Bayreuth Dragoon Regiment and the Gondrecourt Cuirassier Regiments, two of the best regiments in the army. This reinforcement, having joined Count von Palfy, this general moved closer to Karlowitz, but he had barely arrived at the chapel, which is close to the city's fortress, than 70,000 Turkish cavalry fell on him, with a plan to envelop him. He withstood their efforts for more than four hours with all valor imaginable and he reached, with much difficulty, the defile near Peterwardein, where he quickly withstood yet another attack. Finally, he made his withdrawal in good order, not having had in such a sharp and unequal action more than 400 men killed. He had two horses successively killed under him and Count von Breuner was taken prisoner. This was the prelude to the war and it was the Turks who had begun hostilities. It was remarked, as an

---

[3] This is the name that the Turks gave Prince Eugene.

unusual thing that the first action that started the war had occurred in the same place where it had occurred 17 years earlier.

After the return of Count de Palfy, Prince Eugene held a council of war near Futack, a bit below Peterwardein, on the opposite bank of the Danube. There it was said, since the Grand Vizier had crossed the Save, and he had expressed so much desire to engage the Christians, it was to the honor of this general and that of the arms of His Imperial Majesty, in particular, to not avoid battle, but to cross the Danube and move before the Grand Vizier. The council resolved on this and all of the Imperial Army assembled to cross the river. Prince Eugene sent the order to Prince Alexander von Württemberg, who was encamped near Szegedin with a corps of 14,000 men, to support the new fortifications that had been constructed before this fortress, to leave this position and join the army with his troops, which he executed with all diligence.

On 2 August, Eugene gave his orders in writing, both on that which regarded the passage of the troops, as for the manner to position them. The infantry approached the bridges which were over the Danube vis-à-vis Peterwardein. There were two in very good state. The infantry was preparing to cross when the Turks set adrift some floating mills, sending them down stream without the Imperials being able to do anything to stop them. Five boats of the first were carried away and 18 of the second. The damage, as great as it was, only delayed the troops an hour and a half. The care and activity of Count von Löffelholz repaired them as promptly as possible; and finally, all of the army had crossed the Danube except the cavalry reserve, which could not be ready so early and which only crossed that evening. The infantry advanced beyond Peterwardein, into the camp that had been marked out for it, which was covered by some old entrenchments made during the previous year.

Peterwardein was a very good fortress. The new works that had been constructed were defended by trenches called "Caprara." There were two principal trenches, one facing the countryside and another more to the rear, intended to second and sustain them. Two other lesser entrenchments were to the right and left, enclosing a common space, and in the form of a long square. In addition, they were covered by good parapets, wide and deep ditches, and redoubts, but at present only vestiges of them remain, as time has destroyed them. The situation was, all the same, always advantageous. On the right side the slope was very steep, below which ran a road which occupied all of the width from the mountain to the Danube. On the left side was another slope, but gentler; then a spacious valley that ended in a swamp and the swamps of the Danube. The entrenchment was no longer limited by the height on that side. The Imperials had left a sufficient space there to march a few battalions.

To the measure that the Imperials occupied this camp the Turks advanced on them. They encamped during the evening of the 3rd, a league from the Imperial camp, before which they started opening an entrenchment in two locations, and to lay out parallels. This was the Turkish custom of advancing against an enemy who had entrenched himself a bit, by approaches and in the same manner as which one besieges a city in war. This custom is particular to the Turks and I am unaware of any other example of this, which can be found among the ancients or even the modern military annals. Be that what it may, the Turks continued their work aggressively during the night and the next day their lines were pushed to within 50 paces of the entrenchments. They also laid out a parallel, where they raised good batteries of cannon and mortars, and where they placed the elite of their infantry. They began, at 8:00 a.m., to direct a continual fire on the Imperials with their cannon and muskets. The Imperials responded with a few field guns posted at the head of the entrenchment, which did not do much damage.

Prince Eugene seeing his camp besieged, did not think he should await the Turks in his entrenchments; this maneuver would appear too timid, and His Highness was not in the habit of receiving battle. To the contrary, he subscribed to the maxim that he should begin the battle, as luck ordinarily declared itself for the audacious. With this thought in mind, Eugene made his dispositions for the attack in the following order:

Prince Eugene was in command, with Count von Palfy commanding the cavalry and Count von Heister commanding the infantry.

The generals of cavalry on the left wing were Mssrs. The Count von Mercy, Falckenstien, Graven, Veterani, Hamilton, and Prince von Lobkowitz.

In the battle corps were Generals of Artillery the Counts Maximilian von Stahremberg, von Régal, Prince Alexander von Württemberg, Counts von Wallis, von Dann, Ahumada, Leimbrück, Luigenstein, and Marsilly.

In the right wing of this line were the Generals of Cavalry von Falckenstien, Debergéni, Hochberg, Croix, Hauben, d'Eck, and Cordoua.

The troops of this line consisted of the Rabutin Dragoon Regiment (7 sqns), the Batté Dragoon Regiment (7 sqns), Palfy Regiment (7 sqns), Merci Regiment (7 sqns), and the Martigni Regiment (7 sqns) – a total of 42 squadrons.

One infantry force consisted of the Heister Regiment (3 bns), Palfy (2 bns), Alt-Württemberg (3 bns), Haffling (1 bn), and Alexandre von Württemberg (3 bns). Total 12 battalions.

Another consisted of Alt-Daun Infantry Regiment (3 bns), Neuperg (2 bns), Wetzel (3 bns), Régal (3 bns), and Durlach (2 bns). Total 13 battalions.

The third consisted of the Maximilien von Stahremberg Infantry Regiment (2 bns), Bagni (3 bns), Gelschwind (2 bns), Gui von Stahremberg (3 bns), and von Sickingen (1 bn). Total 11 battalions.

The right-wing cavalry consisted of the Graven Regiment (7 sqns), von Falckenstein (7 sqns), Althan Dragoon Regiment (7 sqns), and Bayreuth Dragoons (7 sqns). Total 42 squadrons.

The generals of cavalry of the left of the second line were Mssrs. Batté, Count Nadasdy, Viard, Gondrecourt, Count von Jörger, and de Galbes.

The Artillery Generals were Mssrs. Prince von Bevern, Count von Harrach, Chevalier of the Teutonic Order, Prince Frederick von Württemberg, von Diesbach-Wallis, and the Duke of Aremberg.

The troops of this line were: von Schönborn Dragoon Regiment (7 sqns), Saint-Amour Regiment (7 sqns), de la Croix Regiment (7 sqns), Hausois Regiment (7 sqns), Gondrecourt Regiment (7 sqns), and Cordoua Regiment (5 sqns). Total 40 sqnadrons.

The von Harrach Infantry Regiment (3 bns), Ahumada (1 bn), Bonneval (1 bn), Prinz Frederick von Württemberg (2 bns), Wallis (2 bns), Faber (1 bn), Trautsohn (2 bns), and Tran-Loréna (2 bns). Total 14 battalions.

The Leopold Lorena Infantry Regiment (2 bns), Alcaudete (1 bn), Marsilly (1 bn), Guellen (2 bns), Johann Daun (2 bns), Lancken (2 bns), and Bevern (2 bns). Total 12 battalions.

The right-wing cavalry consisted of the Vasques Regiment (5 sqns), Prinz Emanuel von Savoyen (7 sqns), Viard (7 sqns), Lobkowitz (7 sqns), Jörger Dragoon Regiment (7 sqns), and Galbes Dragoon Regiment (5 sqns). Total 38 squadrons.

The reserve was commanded by General Spleny.

From this one can see that the first line had 84 squadrons and 36 battalions; the second had 78 squadrons and 26 battalions, and the reserve corps contained 26 hussar squadrons: Ebergeny Hussar Regiment (5 sqns) Spleny Hussars (5 sqns), Esterhazy Hussars (6 sqns), Baboczay Hussars (5 sqns), and Nadasdy Hussars (5 sqns).

The total army had 187 squadrons: 56 dragoons, 106 cuirassiers or cavalry, and 25 hussars. It had 62 battalions, half infantry and half cavalry.

All of this army, arrayed in battle order, extended a league, of which the entrenchment covered a bit more than half. The cavalry on the left was covered by a swamp and that of the right by a steep slope. It was always Eugene's first concern to cover his flanks, above all when fighting the Turks, who were ordinarily more numerous than the Imperials, and could turn them and attack their flanks. This is not an insignificant event in a ranged battle.

The Turks were warned of Eugene's resolution to attack them and deployed to receive him. They put themselves in movement on all sides. The sides and valleys were covered with their troops. The Turks numbered 250,000 men, of which 40,000 were Janissaries and 3,000 were Spahis. The remainder consisted of Tartars, Wallachians, Arnautes, and troops from Asia and Egypt, or other similar troops.

From this one can see that they formed a front that was much larger than that of the Imperials; but their line was far less regular, because the barbarians were almost unaware of the art of fighting in rank and file.[4] The power of the Turks came from their numbers and the weight of their charges. They posted their cavalry opposite that of the Emperor. Their intervals were filled with Janissaries, and the remainder of this corps formed behind an oblique valley, where they could easily support their comrades. Another great corps appeared a bit further away on the left, but it remained immobile throughout the battle for no reason that we have yet determined. Perhaps it was a reserve, which had been forgotten in the heat of the action, and no orders were given to it, and it would do nothing without orders. As to the artillery, although the Turks had a great deal of it, it did them little service in the battle, both because it was heavy and it was not easy to bring it forward, as well as there was inadequate time to bring it forward. There were, as a result, only three batteries of cannon, one aimed at the left entrenchments, the other against the center, and the third against the right flank, with a pit of four mortars.

It was 1 August 1716 as the two armies began preparing for what would become the battle of Peterwardein. The two armies took three or four hours to arrange themselves. It was about 7:00 a.m., when Prince Eugene sounded the charge. Prince Alexander von Württemberg began with his brigade, which contained six battalions. He pierced the Turks and penetrated as far as the artillery battery, which they then captured. The Imperial cavalry charged with the same success. The Imperial forces had already declared victory and they began to congratulate each other for the little effusion of blood, when they suddenly saw that the infantry of the right had broken, and that this would produce the opposite effect. I wish to speak of the entrenchments from where

---

[4] It does not appear by all of that has passed during the war which was actually fought in Hungary, that Pasha de Bonneeval had made the Turks great tacticians.

this infantry had to come out of to approach the Turks. Although quite ruined, these trenches were not easy to cross frontally. It was necessary to pass through eight openings, which required the formation of eight columns. Each column was led by a major general of infantry, or by a lieutenant of the Field Marshal. The order was to extend once one was outside of the lines, but the lack of space between the entrenchments and the Turks did not permit this. The Imperials found themselves under the Turks' fire and they had no sooner seen the heads of the columns than they came out of their holes with terrible cries. They were not badly received. The German infantry withstood their shock with an extraordinary vigor, drove them back, and advanced, gaining 20 paces of ground on them. However, this advantage only lasted an instant. The entire corps of Janissaries posted in the valley, fell on the German infantry at a rapid pace. The columns, which had only half passed through the gaps, could not resist such a sharp and heavy charge; and the barbarians profited from their trouble, driving them back and throwing them back on one another. They penetrated as far as the first entrenchment, then pushed to the second. Lieutenant Field Marshals de Bonneval, Lanken, and Wellenstein attempted to re-establish order; but their prayers, exhortations, and threats were useless. Their soldiers were deaf to their voices and disorder was augmented. The Turks continued to saber all that they encountered. Lanken and Wellenstein fell dead as they attempted to rally their troops. Count de Bonneval found himself separated from the column he commanded and in the middle of the Turks, only having around him 200 soldiers of his regiment. Nonetheless, he was not disconcerted and applied his valor and experience, deploying his men hastily to the rear of the same works as the Turks, and arranged them such that they faced in all directions. This little troop defended itself for a half hour; but the 200 were reduced to 25 and it was time to cede. Count de Bonneval thought of retreating. He led his 25 men and cut his way through the body of the Janissaries. This was not without giving and receiving more blows. His little troop was, so to say, slowly being cut down, ten further soldiers being cut down and he was struck by a lance which knocked him to the ground. However, he immediately got up and ran his sword through the Turk who had wounded him. He then retired towards the river.

While the Turks fought the Imperial infantry on the right, their cavalry tore at the Turks. The Spahis charged with great cries, leaping and prancing, but the German squadrons stood like walls, marching at a grave and regulated pace. They soon drove back the Turkish cavalry, making themselves masters of their ground and holding it, though the Turks charged repeatedly. The brigade of Prince von Württemberg maintained itself as well. The reserve was not shaken, and the flanks were guarded. The punishment suffered was not without remedy. The Turks, too dazzled by this ray of victory, did not look to assure that they were not taken in the flank by the Imperials and this flank, which was too long and, in the air, was pierced at the first shock. However, if they did not pay attention to this, Prince Eugene did. His military eye was better that most, so he soon observed the Turk's error. He profited from it skillfully and a promptness that was natural to him. He sent orders to Count von Palfy to detach 2,000 cavalry from the left to pass to the right, and charge in the flanks the Janissaries occupied in forcing the second entrenchment, behind which half of the Imperial infantry, that had been broken, had sought refuge, and where, truly, it could not resist long against such a great number of Turks. The order was well executed. The 2,000 German horsemen had soon pierced the floating and open battalions of Janissaries. They were crushed under the horses' hooves. They were driven back in their turn. This success gave the Imperial infantry of the first and second lines time to re-establish themselves. The battalions reformed themselves and returned to the line. The reserve corps advanced, the artillery of the fortress fired on the Turks. This put them between three or four fires. They did not know where to turn. The art of forming a battalion square was unknown to them. If they had known it then, they might have been able to make an honorable retreat. Seeing no option but flight, they embraced it without hesitation, and they ran in every direction. Most directed themselves towards their works and plunged into their trenches. However, most of the holes that had been dug to save their lives served as their tombs. Death pursued them with blows of swords and bayonets. They might have been able to rally, under the cover of their trenches, and held out for a long time, but such is the nature of the Turkish soldier to not act except by audacity or consternation. This first movement is caused by their impetuosity in the first attacks, which is without a doubt also the effect of their presumptuousness, and the other is the result of their lack of training and their ignorance of the profession of arms. Whatever it might be, their rout was complete. They abandoned, as was their custom, their artillery, ammunition, tents, and baggage. The Imperials did not amuse themselves with pursuing them. They were still in such a great number that it was dangerous to pursue them. Their cavalry was posted in an extremely advantageous position; since that of the Imperials could not go against them, except by moving across ravines and brush, but such is the method of this cavalry, otherwise excellent when on horse, only fighting in a caracole without observing either rank or file, and without making any concerted or unified movement, nor remaining concentrated. This is why they cannot withstand the effect of the German squadrons, accustomed to moving in concert and fighting in dense formations.

The battle had not lasted more than five hours. At noon Prince Eugene entered the Grand Vizier's tent, which was of an extraordinary size and magnificence. He made a short prayer while in the tent, to thank God for the victory He had accorded him. All of the army knelt on the ground to pray on the battlefield, covered with dead Turks, where the army's priests tended to the dead and wounded.

Prince Eugene, having rendered homage to the God of battles, occupied himself with writing a letter to the Emperor to give him word of the victory his troops had won. He also gathered up as many flags and horse tails as possible and sent them to the Emperor with his letter by Count Karl von Zeil, Captain of Dragoons of the Eugene Dragoon Regiment, who had the duty of informing the Emperor of the many particulars that Prince Eugene had not had time to write in his letter, which contained only a few lines.

The number of dead Turks was not known, but it certainly passed 6,000, while those of the Imperials were about 3,000 dead and 2,000 wounded.[5]

The Royal booty consisted of a prodigious quantity of bombs, shot, powder and grenades; 164 cannon and mortars, plus a large number of smaller pieces. The Imperials gathered up 150 flags and standards, five horse tails, and two pairs of kettle drums. All of this was taken to Vienna and placed in Stephansdom Cathedral.

Prince Eugene retained the Grand Vizier's tent. Everything else was abandoned to the Imperial soldiers, who stuffed their pockets and sacks with the riches of Asia. "It is certain," says an author[6], "that if these things (the soldier's booty) had been sold at their true value, they would have all had lives of leisure. However, I know that war booty profits no one. It is dissipated, it is destroyed, and no one knows where it goes."

The news of this victory spread joy throughout Christianity. The Pope gave public expressions of his joy, between others sending to Prince Eugene the present that the Roman Pontiffs had given to various great men who had distinguished themselves against the Muslims. Among these heroes were the Emperors Frederick IV, Maximilian I, Charles V, Ferdinand I, and several other kings and princes. This present consisted of a sword (glaive) named "Estoc" and a hat. The Pope sent these things to Prince Eugene, writing him a letter filled with gracious expressions and praise.

Meanwhile, Prince Eugene was camped on the battlefield, occupied with burying his dead and helping the wounded. His heart, after the victory, showed its tenderness and natural compassion, which did not permit him to be still when it came to the fate of the unfortunate. He could not give life to the dead but attempted to save that of the wounded. He did not spare his purse, nor his credit to hasten their recovery, and in procuring for

---

[5] Translator: The *Nouveau dictionnaire historique des sieges et batailles memorables, et des combats maritimes les plus fameux,* (Paris, Gilbert et Cie, 1809), Vol. 5, p. 176, claims that the Ottomans lost 30,000 soldiers. It also states that the Imperials captured 156 cannon, 172 flags or standards, 5 horse tails, 3 pairs of kettle drums, and all of the provisions of their camp.

[6] Dumont, *Histoire militarie du Prince Eugène*, p. 109.

them all relief available. He had a *Te Deum* sung in his camp, on 8 August. After this was done, he recrossed the Danube to avoid the infection caused by so many dead bodies.

After the defeat of the Janissaries the Grand Vizier had rallied 2,000 cavalry of his guard, which he passed through a defile to charge the Imperials who were pushing the fugitives, but this general, having been abandoned by part of his men, received two wounds from which he died the next day at Karlowitz. An hour before his death he gave cruel marks of his hatred for Christians, by ordering the murder of Count von Breuner, "Finally," he said, "this dog will not follow me: And pray to God that I can exterminate with him all of the infidels." Count von Breuner was greatly regretted. He was a young lord of great hope. His family had volunteered the greater part of their goods to save him.

Prince Eugene, wishing to profit from his victory and the consternation of the Turks, resolved to form the siege of Temesvar. To this end, he detached 16 cavalry regiments, under the orders of Count von Palfy, and 10 battalions, commanded by Prince Alexander von Württemberg, to invest the city, to follow with the rest of his army. Count von Palfy led with his cavalry, with the idea of crossing the Theisse at Salbia, but the waters were too high, which obliged him to move up to Zenta, where he crossed without difficulty, as much as it was possible, his troops not being sufficiently numerous to occupy all of this region. He was obliged, in approaching, to engage a corps of Spahis, which had come out to dispute his crossing of the Theisse, but they arrived too late. They were charged and put to flight.

Temesvar is a strong fortress by its natural position, as well as by the steps taken to fortify it. The Temes, from which the city appears to draw its name, does not flow past it; however, the Beja River does, and it was a branch of the Temes. The low ground which one encounters there stops it. It leaves there by various canals and it forms a large swamp, much of which is flooded. It is in the middle of this swamp where Temesvar was built. The fortress is inaccessible by the ordinary means of trenches from the eastern side, and from the west and south it is barely more accessible. It is only on the northern side where one can approach it over firm ground running 500 to 600 toises[7], but it is frequently unusable outside of the summer, when it is ordinarily flooded.

The entire fortress can be divided into three parts: the city, the castle, and the stockade. The latter covers a suburb called, the "Stockade Suburb." It contains more inhabitants than all of Temesvar, and the stockade is a fortified work with a ditch, constructed in the Turkish fashion, riveted with rocks. The city was fortified with a bit more regularity. It had a good exterior, a covered road, a forward ditch filled with water, with a berm in the ditch, and a good thick rampart. The exterior of the city was not covered with stone, like the Stockade Suburb, but with large logs of 15-18 feet in length so deeply driven into the ground that they only extended seven feet above the ground, forming an excellent palisade. The castle was fortified in the same manner. It stood behind the city and was defended by a smaller stockade.

Prince Eugene, who had stopped a few days on the banks of the Danube on the side of Futack to rest his troops, finally resumed his march and as the infantry could only move slowly because of the extreme heat, he placed the cavalry in the lead and advanced to Zona, where he once again rested his troops. The infantry joined him there. On 25 August, Eugene, followed by his dragoon regiment and the Württemberg Dragoons arrived at the camp near Temesvar. He was joined there next day by all of the army. Upon arriving the troops occupied the different posts that had been designated for them. Once that was done, Prince Eugene began work on the bridges over the swamps to permit communications from the quarters. He also attacked a pleasure house of the city's commanding pasha, situated in one of the suburbs. The Turks did not defend that quarter, judging that they would not hold it very long. They abandoned it, but not before setting it afire, consuming the suburbs and the magazines of forage that had been amassed there with great effort and expense.

On 29 August, Prince Eugene had 30 grenadiers attack a mosque below the stockade. It was carried in an instant, not being defended. The Turks preferred to lose the mosque than to profane it with blood. The Imperials then posted a full company in it.

---

[7] Translator: A toise is 39 English inches, or about a meter.

After the Imperials had carefully examined the ground and taken all of the necessary measures for the success of the attacks, they prepared to open the trench. The opening occurred during the night of 1/2 September to the left of the mosque about 400 paces from the stockade. Two attacks were formed, one of which was pushed on the right hand of the Forforos Gate and the other on the left towards the Mortoros Gate. Prince Alexander von Württemberg commanded the troops of the detachment. He had under his orders Count d'Ahumada, a Spaniard and maréchal de camp, and Duke d'Aremberg, Sergeant General. The next day the Imperial heavy artillery, which had been long awaited, arrived from Peterswardein.

That same day Count Maximilian von Stahremberg relieved the trench with Count Wallis, Maréchal de camp, and Marquis de Marsilly, Sergeant General. The works were pushed ahead and the workers and troops that supported them, began to be under cover from the fortress' cannon. That did not prevent Prince Emanuel of Portugal from going too far forward and, inevitably, he was injured. This young prince had left France in disguise to join the Imperials at the beginning of the campaign. He had given many proofs of his bravery, but this cost him his life. He wished to go into the trench to see what was happening there. He had not asked permission from Prince Eugene, as he anticipated his request would be denied. He went there on his own accord and had pushed his horse towards a spot where he thought he could see some Turks, when a cannon ball from the stockade struck him between the knee pad of his boot and the side of his horse. The animal was killed, and the Prince of Portugal had his knee brushed such that he could not get up. He was carried away and a few days later he came down with a fever, such that one feared for his life. However, he recovered, to the great happiness of the Imperial Court, which had been very alarmed by his wound.

On 3 September, a parallel was pushed 320 paces to the left and a redoubt, with a *place d'arms*, was raised at its head. At the same time, the Imperials began working on two 18-gun batteries. On that day, only four soldiers were killed and 30 wounded, including a captain and a lieutenant.

The batteries began to fire with much success. The same day the Prince of Portugal, who began to recover, wished to go to the trench, despite everything that was said to stop him. Prince Eugene was obliged to expressly prohibit him from going and the Prince did not go.

On the 7th, the works began to be perfected, and there was a 220-pace line drawn for the communications from the second parallel.

On the 8th, it was learned that 13,000 to 14,000 Tartars had crossed the Danube on 50 boats near Panczowa, to ravage the region beyond the Temes, and they had orders to attack any Christian foragers that they encountered.

On the 6th, the garrison executed a sortie (their first), which did little. It was supported by continuous fire from the fortress. The detachment from the garrison attacked the workers, sabers in hand. Several soldiers of the detachment carried burning torches, to set fire to the fascines and gun carriages, but they did not succeed, and they were openly pushed back to the fortress.

Nothing of significance occurred on the 19th, other than the batteries being perfected and the trenches pushed to within 30 paces of the stockade, and the Imperials began battering the walls to create a breach. Count von Harrach entered the trench, directing the perfection of the works and assuring the lodgments that had been begun near the ditch. The Commander of Engineers was killed as the saps were opened in two places. In the meantime, Prince Eugene received word that Mr. Fleischmann, the Resident of the Emperor in Constantinople, who the Turks had poisoned at Semendris when he passed through there to return to Vienna, had been transported to Belgrade where he was locked up.

On the 20th, Count von Steinville, who Prince Eugene had recalled from Transylvania, arrived at the camp before Temesvar, with two battalions from the Wirmont Regiment, one from the Brown Regiment, and one from the Count Ottokar von Stahremberg Regiment, plus four companies of grenadiers, and the Steinville and Neuburg Cuirassier Regiments.

During the evening of the 22nd, Count von Palfy informed Prince Eugene that the hussars that he had openly sent out, reported that the Turkish Army was advancing and that a vanguard had pushed a part of their troops, which served as an escort to the soldiers who worked on the fascines, and that they were marching to attack the Imperials.

Their plan was to relieve the fortress and for that they had resolved to attack during the night of the 23rd/24th, sending a relief column of 12,000, both Spahis and Tartars, into the fortress. Part of them were to carry, behind their saddles, 500-600 picked Janissaries, while the others carried sacks of gunpowder, rice, flour, biscuits, and other provisions, which they knew that the garrison lacked. The Serasker of Belgrade, to favor the entrance of the relief column, had sent a detachment of 2,000 Turks and 8,000 Tartars, which was to force the quarters of Count von Palfy, while those in the city executed a sortie to second them.

Prince Eugene, informed of the Turks' plan, moved to the quarters of Count von Palfy at nightfall, took command the brigade of Count Maximilian von Stahremberg, which contained 11 battalions, and took 24 cannon loaded with canister with him. If Prince Eugene wished to take other precautions, there was no time. A half hour later the Turks arrived, and attacked this quarter with great cries, as was their custom. The Imperial cavalry, which had been posted along the lines of circumvallation, withstood their first shock with an extraordinary vigor and drove the Turks back. Twice more the Turks attacked and both times they were repulsed. They made one last effort to at least introduce a few hundred Janissaries into the fortress, but they did not succeed, and the Imperial cannons, loaded with canister, ravaged them terribly. The garrison had not budged during the entire action. It was claimed that they had been given a specific time and that the relief column had attacked too early. As a result, the garrison did nothing to support it.

Judging by the number of dead that the Turks left on the ground, their losses were considerable and came to about 4,000 men, among which there were many officers, who were easily recognized by the richness of their clothing. Spies and prisoners confirmed that the corps contained 27,000 to 28,000 Turks or Tartars, and according to all information, it contained the best Janissaries of all of the troops of the Grand Seigneur. However, there were only 600 Janissaries they had suffered the greatest casualties because of their great efforts to penetrate into the fortress. The garrison executed a sortie with infantry and cavalry, but it was too late, as the detachment of the Turkish army had retired, and the garrison was obliged to do the same.

On 25 September, Prince von Bevern was on duty, when the works were pushed vigorously in the galleries and on the bridges for the ditch around the Stockade Suburb, where Mr. Mischner, Captain of Engineers, was killed. That day all of the dispositions necessary for the assault were completed and readied for the morning. This was the third or fourth time that those preparations had been made, however, a terrible fire by the garrison made them useless. Bombs rolled out of the stockade on planks ruined the galleries and filled them with water from the Beja, which interrupted all of the works. The attack was delayed. The day was set for 30 September. Prince von Württemberg was in charge of the trench that day and was given command of the attack. He had under his orders Mssrs. D'Ahumada and Leinbrück, with three major generals (Mssrs. Langlet, Liebenstein, and Wallis). He had 3,000 grenadiers supported by 30 battalions and 2,700 pioneers. This force was organized into three corps destined to launch the attack in three locations at the same time.

Prince Eugene ordered Count de Palsy to launch a false attack against the small stockade, which covered the castle, in order to draw the Turks to that side, and weaken them to the real attacks. The action was not launched that day, there not being sufficient time to organize the troops. As a result, the troops spent the night under arms in the approaches, despite a heavy rain that lasted all night. That evening Count von Hohberg, a general officer in the service of the Emperor, was killed by a cannon shot, as was Mr. Harcourt, an ensign in the Brown Regiment. During the morning of 1 October, Prince Eugene came to the approaches to inspect the troops he had ordered for the assault. He gave them some liberties. He had a conference of three quarters of an hour with Prince von Württemberg, General of the Day, after which His Highness of Savoy came to stand by a battery where he could watch what passed during the action.

The signal for the assault was a general discharge of all of the batteries in the approaches. Upon this the troops rose up out of the trenches and began the attack with an incredible fury and valor. The grenadiers were the first to come out of the galleries and advanced over the bridges that had been thrown over the ditches. With great intrepidity, part in the galleries and part on the ditch, they captured the parapet and began a relentless battle where a thousand proofs of their valor were shown. They made a lodgment on the left and chased out the Turks, who hastily fell back to the city fearing that they might be cut off by the Imperials.

At the same time, a battalion posted in the area that was abandoned occupied the Stockade Suburb. It immediately constructed strong entrenchments and had fortified itself so well that the Turks that rallied and returned at the charge to recover this post were vigorously received, pushed back, and again driven back to the city. Their only success in this attack was setting fires in some parts of the suburb. This did not prevent the other battalions from taking post in the suburb, and while taking advantage of the works that the Turks had constructed 80 to 100 paces from the fortress, and where they began to dig a parallel line at the same distance from the ditch of the city. The Imperials improved the captured Turkish works as best they could and, with great diligence, the lodgments that they had constructed.

This attack required all the valor and intrepidity of the officers and the German soldiers to surmount the inflexible firmness of the Turks, who were redoubtable in the defense of a breach because of their dexterity with the saber and the idea that they must never surrender any fortress.

This action lasted nearly four hours and it was most bloody. Initially the Imperial losses were estimated to be 500 to 600 dead and wounded, but it soon rose to 1,000 dead and wounded. In the following days, an exact number was determined, and the Imperials had lost 1,327 wounded, and more than 400 dead. In addition, they had lost 33 captains, 52 lieutenants, and 123 non-commissioned officers killed. Count Maximilian von Stahremberg was wounded. Prince Alexander von Württemberg, who commanded the attack, received a contusion that was not significant. Mssrs. D'Ahumada, Brown, Liebenstein, Tattenbach, Faber, Rudolfing, Geier, Baron Kazianger, Fack, Dégano, Consoda, Wisse, Hochbarth, Pfeffershofen, Dietrich, Somoviva, and Count

Hamilton were wounded. Tattenbach and Kazianger died from their wounds. Baron von Beck was the only officer of distinction who was killed in the battle. In the suburb, the Imperials found a considerable quantity of cattle and horses, though many had died in the fires started by the Turks, and by the fires that the Imperials subsequently set in certain houses or other places where detachments of the garrison had entrenched themselves. The fires continued for a further eight or ten days and it was estimated that 1,200 houses had been destroyed. This was about half of those that were in the suburb.

Prince Eugene sent a courier to Vienna the day that the Stockade Suburb was captured. He reached Vienna on the 5th and the news he carried spread joy throughout the Court and city as they knew that once that suburb had fallen the city would follow soon.

On 3 October, the work on the trenches continued such that the parallel was close to one of the city's demi-lunes. The fire was very violent by both sides and above all from the Imperials.

On the 4th, a second parallel was pushed to the left of the Stockade Suburb a further 260 paces and it was pushed as far as the swamp. Prince Eugene had a battery constructed on the side of the attack. It contained 15 heavy cannon and seven mortars on a platform which had been raised the previous night with great risk because the workers were exposed to the fire of the Turkish cannon in the fortress while they worked.

The next day was spent in perfecting the batteries and that same day a prisoner who had escaped from the Tartars arrived in the camp. He informed Prince Eugene that a day and a half from Temesvar there was a Tartar camp, and closer to the Danube there was a considerable corps of Turks. He also said that during the attack in the quarter of Count Palfy, when the Turks had attempted to throw a relief column into the city, the Aga of the Janissaries, who was then called the Serasker of Belgrade, had given a golden sequin (golden ducat) to each of the soldiers that had passed beyond the Danube in order to give them more courage and motivation.

The siege work was pushed hard on the 6th, and the Imperials began to fire mortar bombs into the city with 14 mortars, causing much damage and disorder.

On the 7th, the Imperials built a redoubt to the right of the attack to protect the batteries of cannon and mortars that were firing on the city and a new battery was begun in the Stockade Suburb, to dismount the Turkish guns. A lodgment was perfected on its left, 50 paces from the ditch, and the rest of the mortars were positioned. There were now 30 mortars continuously firing into the city.

The next day the Imperials pursued the works with force, to put the batteries in a state such that they could batter a breach in the body of the fortress the next day or the day after.

On the 11th, the Imperials began to fire on the fortress with 50 cannon and the mortars continued to fire in a horrible manner. The Turks responded only weakly to this horrible fire, because of the disorder that the Imperial fire had caused in their batteries. However, on the 12th, their batteries were repaired, and they then began a terrible cannon and musketry fire that lasted until nightfall. That produced concern in the Imperial ranks that despite the cares and efforts of Prince Eugene; the skill and bravery of his soldiers, the enterprise would be checked by bad weather and continual rain which fell and filled the trenches with water. It was thought that if the Turks held out a few more days the season would be too advanced, and the besiegers would be forced to withdraw. Prince Eugene was beginning to believe he would have to withdraw, when he saw, during the morning of the 13th, a white flag hanging over the fortress' works. Prince Alexander von Württemberg, who was on duty that day, hastened to bring the news before he was relieved. Prince Eugene, learning of this, consented that the Pasha would send a few officers to his camp and offered, on his side, to send into the city some of his own to serve as hostages. Everything being executed, the hostages were exchanged, and the following articles of capitulation were agreed upon:

I.   That the Turks may leave Temesvar with their women and children, their horses, and their cattle, with the wagons necessary to transport all of their goods that are in their houses and that all of it shall remain at their disposition; that one would grant them the liberty to leave; that one would not hinder them in any way.

      This article was accorded in its totality; but as the besiegers had not added anything regarding deserters, they were excepted from those who came out:

II.   That the soldiers, or militia, on foot or mounted, and the inhabitants, may leave with their weapons, blades or firearms, their flags deployed, drums and kettle drums sounding: That they shall be led, in seven or eight days' march, including that of the exit from the city, to Borscha, near Belgrade, by the shortest path, and with a sufficient escort: That the first day they shall move towards Temesch, above the bridge; the 2nd by the second bridge, near the Schebel, which is a village in the swamp, the 3rd to Tente on the bridge over the Bieschowa; the 4th to Margida; the 5th to Allibonar, near a stockade; the 6th to Panczowa; the 7th at Borscha, which is the objective.

      This article was granted under the condition that the garrison would leave hostages for the security of the garrison, and that on the last day they should be given a certificate signed by the Pasha of Belgrade, that they had delivered the garrison to Borscha; that finally, the hostages should be released upon the return of the escort.

III.   That the Imperials shall furnish 7,000 wagons for the transport of the women, children, effects, and merchandise, and in the case where one is broken or the draft animals may die on the road from fatigue, it shall be replaced. That one shall permit individuals to purchase, with their money, any wagons that they find.

Prince Eugene only wished to give them 1,000 wagons, it being impossible to give them so many as they had requested, but he did permit them to purchase as many as they wanted, and to use those that they already had. He gave them every assurance that they might wish in regard to all that belonged to them; however, he demanded of them that they would not commit any act of hostility or intimidation towards the escort when it returned.

IV.   That the provisions necessary for the subsistence of the garrison, during its march, would be brought to it by the peasants at a reasonable price.

Prince Eugene responded that provisions would be provided to them on the road at a reasonable price.

V.   That the convoy or the escort, during the march from Temesvar to Belgrade, shall not mix with the Turks; but that these troops shall conduct themselves in good order and march on the wings, to put them under cover and assure that the Hungarians, Rasciens, or any other group give them any insult or obstacle.

This article was granted in its totality.

VI.   That after the capitulation was concluded and signed, and when the ammunition, artillery, provisions, and other necessary things were faithfully delivered, one shall reserve to the individual families, to whom shall be granted the right to take with them, without obstacle, what they wish, and which they can carry with them; and that they shall be given the right to dispose of it as they may wish; that with regard to the surrender of the exterior works and a gate, those who are charged with regulating these articles, shall have a full and sufficient power to treat in what manner and when they shall be handed over.

Initially there was some consternation on this article, but eventually it was granted.

VII. That the slaves and all other Christians, who have voluntarily and for a long-time embraced Islam who come out with the others, shall not be detained; with the exception of the defectors and those who have deserted during the siege, who shall be taken into custody where one finds them or they are encountered. That for the same reason, the Rasciens, the Greeks, the Armenians, or people of other nations living in Temesvar, who exercised there their profession, shall have all liberty to remain in the city with all of their effects and shall not be detained, if they wish to leave, and if so, one will not restrain them in any manner.

This article was accorded, on the condition that all of deserters be faithfully surrendered and that they give the greatest assurance to engage the Jews, the Armenians, and others who live in the Banat and Temesvar; and to ensure them complete liberty with regards to their persons and their business, or if they wish to leave, to go wherever they may wish.

VIII. That there shall be equally permitted to the Coruzzes (which were a group of bandits who lived by pillage) who find themselves in Temesvar permission to retire to Belgrade.

The Imperials responded to this article with these few words: "This rabble can retire wherever it wishes."

IX. That the liberty to sell all property be granted, and that in general all of those who leave the city may freely sell their goods and all of their property.

Granted in its entirety.

X. That upon leaving there shall be no impediment, no obstacle, that the capitulation shall not be violated, under the pretext of searches, to pass or for some old grievance.

All this was accorded, and some days alter the city was evacuated.

Temeswar was thus taken 44 days after the trench was opened. The garrison left the city with 12,000 men, not including the sick and wounded; it had started with 18,000 at the beginning of the siege, of which 3,000 were killed. The Imperials lost 4,000 soldiers, but this loss was compensated for by the importance of the conquest and the news of its capture spread joy throughout the Court and all of the Empire. After it was occupied, an inventory was made of the artillery and ammunition in the fortress, and 120 cannon were found bearing the crest of the Emperors of the House of Austria, some of which dated from when the city was captured by the Grand Vizier of Sultan Suleiman II.

After the loss of Temesvar the Turks abandoned various small posts in the vicinity, which Prince Eugene quickly occupied. At the same time, he gave orders to sweep the city, which was filled with the debris of the houses destroyed by the bombardment.

Prince Eugene gave great proofs of his goodness and generosity in all that he did for the troops of the garrison, to whom he accorded things that had not been stipulated in the capitulation. He also took extreme care of the sick and wished to show the Turks the difference between them and Christians in terms of humanity, he did all sorts of kindnesses for the officers and soldiers of the garrison. This conduct was quite the opposite that had been shown in the 16th century by these barbarians when they had captured Temesvar and had given an honorable capitulation to the Christians who had garrisoned the city. The Turks had massacred them once they came out of the city. The Most Serene Prince did not content himself with examples of his goodness, he wished those under him to follow his example. He sent orders to all of the governors and generals who commanded various army corps in different cantons, to treat their prisoners with all gentleness possible.

The capture of Temesvar was followed by a success that was no less considerable, that of the submission of Wallachia to the domination of the Emperor, and the capture of Maurocordato, the Hospodar of this province; a dignity to which the Grand Seigneur had recently raised him and which he did not enjoy very long.

Prince Eugene put a garrison into Temesvar and established Count von Wallis as its interim governor; however, the Emperor wishing to give the Prince marks of his complete confidence, confirmed the choice that he had made. Prince Eugene had also put garrisons in Mehadia, Panczowa, Vipalanka, all important posts; and after having erected the establishment of winter quarters, he prepared to return to Vienna. He learned, in the meantime, that the Chevalier Rasponi had brought him, on behalf of the Pope, the hat and the blessed sword, and that this knight awaited him at Javarin or Raab, to give them to him in his passage. The Prince wished that this ceremony be held in Javarin and having informed General Heister, Field Marshal and Governor of this city, who, despite his great age had been at the battle of Peterwardein, and at the siege of Temesvar, left before Prince Eugene went to Javarin in order to make the necessary preparations.

On 6 November, Prince Eugene left Bude to go to Javarin. He found, in approaching this city, two companies of cavalry well-dressed and very agile, consisting of young gentlemen and the best bourgeois of the city. Once the Berlin carriage of the Prince appeared, the trumpets and kettle drums began to sound, and the two companies formed in two files between which the Prince passed. This troop then broke into four bodies, one galloping before his carriage, two on the side, and the fourth to the rear. The Governor sent a captain of the garrison to meet the Prince and beg him to accept an apartment in the castle. The Prince refused because of the great distance between it and the church. He preferred to take a lodging in a private house much closer to the church. The Prince entered the city by the Stuhlweissenburg Gate where he found a body of bourgeois infantry arranged in a hedge, drums beating, flags flying. All of the garrison was also under arms and arrayed in battle formation on the main square.

Upon entering the city Prince Eugene found Field Marshal Heister, who awaited him with a six-horse carriage. This venerable old man, after having greeted his generalissimo, begged him to accept a place in the carriage in which he rode. The Prince joined General Heister, and the Chevalier Rasponi[8], who had also the honor of greeting Prince Eugene, who gave him a letter from the Pope.

The next day, at 10:00 a.m., the Prince went to the cathedral, where Mr. de Gondor, Bishop and Vicar General of Raab, received him and congratulated him at the head of his clergy. He then led the Prince to a magnificent seat, covered with a canopy shining with gold and gems. The prelate celebrated the Mass, which was formally sung to the continuous discharges of cannon and musketry from the square. The Bishop read the Pope's letter, after which he presented the sword Estoc to the Prince and placed the hat on his head. During this ceremony, Prince Eugene had at his side Prince Emanuel of Portugal and Field Marshal Heister, with several other lords and general officers.

The hat was violet and trimmed with ermine. On the front, there was the Holy Spirit in the form of a dove formed with little pearls artistically placed, and on the two sides there were two golden ribbons, the cord of which was also laced with gold. Above these were three very fine little pearls. The sword was more than four feet long and the handle alone was more than 10 inches long. The guard was of silver, weighing about seven pounds. The blade was two-and-a-half inches wide. The scabbard was of red velour, as was the belt. After the ceremony, the Prince went to dine with the Governor, who served a magnificent dinner, and the food was most delicious. The dessert was most ingenious. It represented all that Eugene had done during the last campaign. The inhabitants of Javarin gave the most extraordinary marks of their affection and respect for the Prince on this occasion. Prince Eugene then wrote a letter to the Pope to thank him for his kind gifts.

In addition, a medal was struck for the occasion, commemorating the victory won by Prince Eugene at Peterwardein.

---

[8] Rasponi had wished to participate in the 1717 Hungarian campaign in the capacity of an aide-de-camp to Prince Eugene. He was killed in a duel by a German officer, with whom he had a dispute, so it was said, over a woman.

End of the 1716 Campaign

Peterwardein Medal.

Frederick Wilhelm Karl von Schmettau

THE 1717 CAMPAIGN

At the end of the preceding year, Prince Eugene had returned to Vienna, where he occupied himself with arranging the finances of the Emperor, in augmenting his funds, and finding other means to sustain the expenses of the war. General Heister had also gone to Vienna to support this effort.

On 3 December, Mr. Fleischmann had arrived in Vienna, the Turks judging it proper to put him at liberty after having detained him in the prisons of Semendria and Belgrade for a long time. This act by the Muslims caused the Imperials to believe that they were looking to arrange a peace, but they were soon disillusioned, when they learned that the Divan had resolved to continue the war and was taking steps to repair the losses suffered in the previous campaign. Fleischmann publicly entered Vienna and was given an audience by Prince Eugene, to whom, as President of the Council of War, he had to report on the negotiations he had undertaken with the Grand Seigneur.

The efforts at recruiting new soldiers proved very successful. Those who enrolled marched with the hope of taking part in the pillaging of the barbarians. Hungary, which the German troops had always regarded as their cemetery, and which they had given that name[9], did not discourage the soldiers who sought service under Eugene. They would have followed him to the edge of the Earth without thinking about the difficulties of the various climates. Uniquely they were occupied with the glory they were sure they would acquire under this hero, and they counted the rest as insignificant.

All of Europe was attentive to the events of this campaign, which was thought to be as successful as the last, which was only its prelude. An extraordinary number of Christian princes and lords came to offer their services to Prince Eugene. One saw everywhere an extreme desire to serve under this great captain, more for the pleasure of being praised one day, than for the interest of religion and the Emperor.

---

[9] It was said that Hungary is the German cemetery, as it was said that Italy was the French tomb.

I will not provide a list of the Imperial regiments, which would become boring. I will content myself with saying that the army that served in Hungary was strong and included auxiliary troops, which came to 140,000 men, of which there were 73,800 infantry, 26,000 cavalry, 15,700 dragoons, 10,000 hussars, 6,000 Rasciens, 3,000 Croats, 750 artillerists, and 4,750 which were expected from the Lowlands. The preparations for food and military provisions were no less extraordinary and the armament of the Danube surpassed what had ever been seen before.

The Imperial cavalry found itself needing a quantity of horses to replace those that had died during the army's stay in winter quarters, or enroute. The famous Jew Oppenheimer offered to furnish, in a few days, all the remounts necessary to put the cavalry in an operational state and to fill, in a short time, all of the magazines of forage and oats to feed these horses. The Imperial Court authorized him to act on all of these things, which were furnished, but Oppenheimer was amply compensated for his efforts, and it was necessary to pay him interest on the sums that he had advanced.

Despite the ravages that the plague had caused in Constantinople, the Ottoman Court pressed its preparations against the Christians. Its fleets and armies were formidable, and it not only continued to threaten the Island of Corfu, where it had been checked the previous year, but it planned to recapture Temesvar.

Prince Eugene spent the winter in Vienna participating in various councils of war with the other ministers of the Emperor, as they prepared plans for the upcoming campaign.

From the beginning of May the auxiliary troops of the various princes of the Empire began to move on Futak. Six thousand Bavarians, commanded by the Lieutenant General, the Marquis de Massei, arrived, on 24 June, at Vienna, and encamped on the banks of the Danube. The next day these troops marched to Château de la Savorite, where the Emperor reviewed them. He found them very fine, especially the 150 Horse Grenadiers, whose blue coats were striped with silver and were most magnificent. The review was attended by the Empress, who expressed much satisfaction at the state of the Bavarians and spoke highly of them to General Massei. The Hessians had already preceded them, under the command of Prince Maximilian of Hesse.

The Electoral Prince and Prince Ferdinand of Bavaria, his brother, had also passed through Hungary en route to Hungary, so that they might serve as volunteers with the Bavarian troops. The reception that the Imperial Court gave these two young princes was most gracious and very different from the treatment they had experienced earlier during their ten years of captivity. The Electoral Prince was particularly in grace with the Emperor, who claimed that a letter written by the young Prince to the Emperor was the cause. The Electoral Prince having recovered his liberty during the time that his father, the Elector, had recovered his states by the Treaties of Rastadt and Baden, wrote to the Emperor in a manner of the world more than spiritual. He thanked him for the care that he had received from him since the Emperor had arrived at the Empire, assuring him that he would have his eternal gratitude.

In addition to the Bavarian and Hessian Princes, there were many other German princes and lords who joined the campaign to serve as volunteers. Of this number were the Princes von Bevern, von Kulmbch, von Württemberg, von Lichtenstein, and von Anhalt-Dessau. The French princes of the blood, Count de Charolois and the Prince of Dombes, animated by the natural noble fire of this nation, also went to Vienna, and on to Hungary. A number of French lords also followed their example: the Prince de Marsillae, of the house of Rochefoucault; the Prince de Pons; the Marquis d'Alincourt, son of Maréchal de Villeroi, celebrated for his many campaigns; the Count d'Estrade, Maréchal de camp in the French Army who served as governor to Prince de Dombes, and several other officers and lords of distinction.

However, Prince Eugene, again named Supreme Commander of the Army of Hungary, left to take up his command in early May. On 2 May, he took his leave from the Emperor. The Emperor, naturally pious and good, after having renewed to Prince Eugene his assurances of the confidence he had in the Prince's zeal, his valor, and his capacity, reminded him that success depended on God, who held in His hands the fate of armies and generals. In saying this, the Emperor presented Prince Eugene with a magnificent crucifix set with diamonds, of a very great value. In presenting it he said, "never forget that [you] go to fight for the interests of He who had died for us on the Cross and to recognize, in this figure that represents Him, the superior auspices under which [you] go to war against the enemy of the name Christian." Eugene received this present with the warmest sentiments of respect and appreciation, and the response he made conformed to the piety of his Emperor.

All preparations made, Prince Eugene left Vienna in the middle of May and at the end of the month he arrived near Futack, then moved to Peterwardein, which is on the other side of the Danube, to meet the troops that had assembled there. From there he returned to Futack, where he remained a few days.

The Viennese Court had resolved on a siege of Belgrade. The Turks had anticipated this city would be the next target of the Christian army. They forgot nothing to put the city in a state to withstand a long siege and to make it impregnable. They employed a multitude of workers and spared neither care nor expense to achieve their goals. They had constructed an entrenchment that ran from the little Gorsca River to the Save River, covering two leagues of ground, and defended it with a ditch that was 18 feet deep. From this ditch to the Danube was a vast ground capable of housing an army of 100,000 men arrayed in battle formation, and everything was prepared so that its flanks, rear, and front were well covered and supported. The two rivers that were there covered the two flanks. Their rear was covered by the great river, and to their front were the entrenchments of which I have spoken, part of which were naturally defended by steep mountains on which the Turks proposed to place their artillery, to put under cover the ground that extended from these mountains to the banks of the Save. All of the exterior of these works was embraced by another ditch six feet wide and 12 feet deep, with palisaded redoubts spaced every 500 paces from one another. All these works were destined to cover the troops that would be posted there to block any movement against the fortress; and the Muslims worked on a third ditch, a short musket shot from the advanced works of the city, which was intended to serve as sort of an advanced ditch.

The works that the Turks had constructed to cover Belgrade naturally leads me to a discussion of this fortress. It is built on the extremity of a hill that forms the junction of the Save and the Danube, and at the summit of this hill is a castle or citadel which commands all of the city. Close to the citadel is a fort known as the "Old Castle," which is defended by two crown works and several modern fortifications, which occupy all of the top of the mountain. The city can be divided into three distinct parts, which are the upper city, the lower city, and the citadel. The lower city is the largest part of the city. It is enclosed by good ramparts flanked by several towers, some being round and others being square. They have a circuit of around 900 toises. On its east, and absolutely outside of its wall, there is a gate that forms a sort of trench, 20 toises wide at its entrance, and about the same as it moves into its basin. The width of the lower city is more than 800 toises. A large tower is attached to its walls, defending its gate, with batteries that can fire on the two rivers; such that one cannot approach by water without being exposed to the most violent artillery fire. Belgrade is a beautiful and well-populated, large city. The streets are narrow, but its houses are extremely full. It is convenient to walk most of the streets sheltered from the weather by means of the trees planted on the right and left, and so thick that neither the sun nor the rain can penetrate to the street. Two great squares, called "Bezestens" are part of the interior ornaments of the city, along with a large number of mosques, the principal of which is attached to a magnificent house known as the "Palace of the Grand Vizier" because, one says, it was a grand vizier who had constructed it.

The city is very mercantile, conducting much business. There are many stores, and all of the streets are full, but many were very short and those who come to purchase merchandise do not know how to enter, because the merchant sits on a bench that closes the entrance to his store. Those who wish to purchase in quantity go to the places where there are large magazines furnished with all sorts of merchandise from Europe, Asia, and the Indies, the transport of which is by the Save, the Danube, the Drave, the Morave, and the Theisse.

In addition, Belgrade had always been regarded by the Christians as a rampart of Christianity and the Turks had made every effort to make themselves master of it, both to cover their frontiers, as well as to open the door for any invasion of Germany when they thought it appropriate.

The despots of Serbia, to whom it belonged, fearing that they could not conserve it from the efforts of the Ottoman Turks, had sold it to the unfortunate Sigismond, Emperor and King of Hungary, who had had constructed most of its fortifications, which were excellent for the time. Armurat II besieged it in 1442 and had been obliged to raise the siege after losing the most part of his army, despite having made a breach with his cannon that fired 100-pound balls.

In 1456 Mahomet II, son of Aumurath, wished to see if he was more fortunate than his father. He put himself at the head of a powerful army supported by a numerous artillery. He covered the Danube with ships to prevent the Christians from relieving the place, where the famous Jean Corvin, better known as Hunniades, Vaivode of Transylvania and Governor of Hungary, had thrown himself and resolved to perish or save this important fortress. He was admirably seconded by the Cordelier[10] that Pope Alexander VIII had sent to Hungary to preach a crusade. This monk knew how to skillfully profit from the idea that the troops had of his saintliness; and his exhortations were always filled with promises of God's forgiveness, the crown of the martyr, and several other things that preachers had used to excite soldiers to their cause. Religion had fought valiantly against those of a different religion, having the effect of giving the garrison a reinforcement equal to 100,000 men. The garrison withstood famine and all of the other inconveniences with an admirable patience and acted in attacks with an extraordinary valor. There came the day when a sortie was led by the Cordelier. The battle was furious and relentlessness. Capistran, crucifix in hand, found himself wherever the danger was greatest, and he obliged the Christian soldiers, by his words to perish before they fell back. He was seen in the middle of the melee, raising his crucifix and crying loudly that those who fought for victory would receive the crown of the martyr. His tranquility, his intrepidity, and more the fortune that he was not wounded, even though he had frequently exposed himself, caused

---

[10] Jean de Capistran was raised to sainthood by Pope Leo.

the garrison to believe that God had made him invulnerable, and perhaps he imagined it himself. However, such miracles were rare. He was enveloped and would have been cut to pieces if Hunniades, who had fought on the other side, after having chased off the Turks, had not come to his succor. The Cordelier was freed, but the battle continued with great fury until the Hungarian Army arrived and obliged Mahomet II to take advantage of the fall night to retreat. Hunniades died of the wounds he had received in the battle and Mahomet II lost an eye.

Belgrade remained peaceful until 1521, when Hungary was torn by troubles and dissensions. Suleiman II, thinking the situation favorable to capture the city, he profited from it and captured the city. It remained in the barbarians' hands until the time that the Elector of Bavaria carried it by assault.

The Turks besieged it again in 1690 under the orders of the Grand Vizier. The Duke de Crois, an old general who had acquired a great reputation, received orders to occupy the city, which he did on 8 October. That same evening a mortar bomb, coming from the Turkish camp, fell on the great tower, set the gunpowder alight, and destroyed the tower with a thunderous explosion. The damage was great. Part of the wall collapsed with the batteries on it, and there was a breach where the Turks could enter by squadrons. Part of the troops, which was the neighboring bodyguard, and most of them who relieved them, were buried under the ruins of the walls and more than 1,000 soldiers who were in the place d'armes were killed or wounded. The fire spread into other magazines and houses suddenly began burning. Magazines began to explode taking with them guns and gunners. The Duke de Crois and the Count d'Apremont were obliged to leap out of the windows of their lodgings and into the street. Repeatedly risking their lives, they rushed to the city's port where some boats transported them to Essek. The Turks profited from the disorder that reigned in the city. They entered it and slaughtered the garrison. Six thousand men of the garrison were put to the sword. A number of inhabitants suffered the same fate, and the Muslims did not cease killing until there was no one left to strike.

Three years later the Imperial Army, under the same Duke de Crois, besieged Belgrade, but in vain. The Grand Vizier had sufficient time to come up with a powerful army and obliged him to raise his siege. The Imperials were obliged to retire after having lost 10,000 to 12,000 men.

The time had come for Prince Eugene to avenge the Christian blood that had been shed below the walls of Belgrade. Several generals of the Imperial Army doubted the enterprise would be successful. It appeared impossible to them that one could surmount the obstacles that they would encounter in the execution of this project, and they were not unaware that the Ottoman Army was assembling below Adrianople, to come to the aide of Belgrade, which the Turks were certain would be attacked. However, Eugene did not allow himself to be stopped by these concerns. He proposed this siege at the Emperor's Council, at which it was approved, directing him to execute the siege. It was to this end that Prince Eugene sent a courier to Vienna to press for the departure of three ships which were to reinforce the naval armament that was already on the Danube. These three ships were the *Saint-Eugene*, the *Saint Marie*, and the *Saint Stephan*. The Turkish fleet was still superior to the Christian fleet and in this manner the Muslims were in a position to prevent the Imperial Army from approaching Belgrade. However, the arrival of the ships mentioned above, changed the state of affairs.

Before undertaking anything against Belgrade, Prince Eugene thought to reinforce the frontiers of the Empire against Turkish raids that might be executed during the siege. He sent orders to Count de Steinville to guard the passes into Transylvania, particularly the one known as the Iron Gate. He charged M. de Viard to cover the Banat or Count de Temesvar, and to have an eye on the bridges that were being thrown over the Danube to communicate with the provinces which are on the right of the river. Prince Eugene did not await the arrival of these new ships before crossing the Danube with his army, in order to invest Belgrade. The Turks took several steps to stop the crossing. The river was covered with their saiques. They had raised batteries, filled with cannon, along the shores, intended to fire on Imperial ships. In a word, the siege of Belgrade became daily more difficult.

After the death of the Grand Vizier, killed in the battle of Peterwardein, the Sultan conferred this most important task on someone who could usefully serve in this critical moment. He had a sad experience with the superiority of the Emperor's generals over his own. He judged that his choice should fall on a captain whose military genius and capacity could balance the talents of Prince Eugene. The Pasha of Belgrade, named Hastchi Ali, was the man selected by the Grand Vizier. He could not have made a better choice. Hastchi Ali had valor, the drive, and the penetration necessary. He had given proofs of these on numerous occasions, and he was superior to the Grand Vizier's choice of the previous year. He made the most judicious dispositions for the conservation of Belgrade. He ordered the assembly of the two corps of troops that were in Hungary. One of them, under the orders of Human Kiuperli, the new Pasha of Bosnia, was to be employed in covering Belgrade, in being posted in the lines that had been constructed before Belgrade and the other was to act in the confines of Wallachia to prevent the Imperials from drawing from their garrisons in Transylvania. Prince Eugene, constantly pursuing his project, ordered on 8 June, that all of the Imperial Army prepare to march out on the 10th.

It was on that day that this army left the vicinity of Futack to move on Belgrade. It contained more than 10,000 of the best men and the most brilliant troops of the Empire. It occupied a camp laid out for it under Titul. The next day it crossed the Theisse and the Beja, over bridges that had been prepared for it. On the 13th, it encamped at Visnitza, near Pančevo. Three warships and some demi-galleys came close to the camp because of a canal that Count von Merci had had constructed in this vicinity during the winter. Prince Eugene had brought these ships up to cover the crossing of the army over a bridge that was to be constructed over the Danube and to hold in respect any Turkish troops that might approach from the far side of the river.

The passage being resolved, Count von Merci, who had 37 battalions and 24 squadrons under his command, had orders to move a league and a half above Pančevo, to embark his corps. The first transport loaded without issues. Prince Eugene ordered the rest of the army to follow. All of the machines and material necessary to construct a bridge were brought forward. As the second embarkation began the Turks appeared, looking as if they wished to charge the troops that were about to be landed, and those who had already landed. The Turks found the Imperials ready to receive them and they did not dare to attack. They withdrew in confusion to the heights around Belgrade. The princes of the blood from France distinguished themselves by the ardor they demonstrated by coming to grips with the Turks. The impatience of Count de Charolois nearly cost him his life. He threw himself into a boat that was overloaded. A rowboat came to save him and drew him to safety. The Bavarian princes and, in general, all of the volunteers of note, both German and French, did much to be noted.

The bridge was successfully constructed using 84 boats, and all of the army completed the crossing despite the Turks, who advanced several times seeking, vainly, to prevent the crossing.

The army encamped at Vinitza heights on the 16th, 2½ leagues below Belgrade, and they laid out a battery of cannon, in the head of the camp, to support six battalions and some cavalry which was under the orders of Count d'Odiver, was intended to cover the bridge.

On the 17th, the baggage began to pass and next all of it arrived in the camp with little loss, although the Tartars and Spahis did not cease seeking to pillage it.

On the 18th, Prince Eugene directed Count von Palfy to lead the troops that were to invest Belgrade. The quartermasters and billeting officers were sent to examine the ground between the Danube and the Save, and to select a camp site. They initially saw that they had a huge extent of ground to cover, because of the entrenchments that the Turks had raised to cover the exterior of the fortress.

Prince Eugene wished to personally examine the situation of the fortress. He was escorted there by six regiments of cavalry, all of the carabiniers, some horse grenadiers of the army, and accompanied by Prince Alexander von Württemberg and several general officers. The weather was not appropriate for laying out a camp, so Prince Eugene returned to camp.

On the 19th, Eugene returned with the same escort to the location that had been designated for the camp. He had barely gone a quarter of a league when he saw a force of 1,200 Turkish cavalry that appeared to wish to engage his escort. Eugene arrayed his force in battle formation and awaited the Muslims with calm. There was a sharp skirmish. The Spahis detached small groups that fell on the Imperials with extraordinary rapidity, and then withdrew.

One of their officers pushed into the second rank of the Imperial cavalry, near Prince Eugene, and offered the butt of his pistol to Eugene. Eugene moved forward to talk to him, but there was not time, as the Turk was cut down by a dozen carbine shots. The Muslims were repulsed, and Prince Eugene marked out his camp in their presence. He sent orders to Count von Palfy to advance the troops that were to form the advance guard. All of the army moved forward and was formed into four columns. It moved to occupy the camp that had been designated to it. As this march could only be made by moving along the Danube, the Turks advanced several saiques[11], which fired vigorously against the Imperial troops. However, several batteries were ordered established on the riverbanks by Prince Eugene, so the ships were obliged to pull away. The army continued its march, and between 9:00 and 10:00 a.m., they began to appear on the Belgrade plain. The left wing extended to the Sava, despite a body of Turkish cavalry which appeared anew and skirmished for more than an hour to block the Imperials' movement. The right wing posted itself, extending to the Danube and deployed new batteries to respond to the Turkish ships that had once again moved up to discommode the army. An Imperial warship, which found itself at the mouth of the Temes, advanced to cut off the Turkish saiques, which it did with some success. Four other warships posted themselves on the other side to intercept any enterprises that those of the city might wish to make by water. Two of these ships, the *St. Charles Borromée* and the *St. Leopold*, which had served at the crossing of the Danube, were left behind between Belgrade and the bridge under the orders of Captain Schindeman,

[11] Another term for a Saic – a type of boat of the Eastern Mediterranean.

and the two others commanded by Captain Storch were posted vis-à-vis Semlin, where a camp of several thousand men, under Field Marshal Count von Hauben, was formed. It was intended to assure the communication with Peterwardein, which was the principal source of their supplies.

Barely had the Emperor's ships rested in the situation mentioned above than they were attacked by five or six Turkish galleys, with more than 40 saiques or demi-galleys. The battle was long and stubborn, the firing lasting for more than two hours. However, the Turks were finally driven back with losses and the Imperials remained masters of the river. The camp found itself surrounded. The city of Belgrade was now invested and enclosed from the Sava to the Danube.

Prince Eugene arranged the post of each general. His Serene Highness, Commander-in-Chief, had under his orders Field Marshals Prince Alexander von Württemberg, Heister, and Palfy.

1st Line

Generals of Cavalry Mssrs. de Montecculi, Montigni, and Ebergeni.

Generals of Infantry: Mssrs. de Régal and Count Maximilian von Stahremberg.

Lieutenant Generals of Cavalry: Mssrs. De Croix, de Hautois, de Bonneval, von Wehlen, and von Walmerode.

Lieutenant Generals of Infantry: Mssrs. Von Daun, Brown, and D'Ahurnada.

Major Generals of Cavalry: Mssrs. Cordoua, Orseti, Marsilly, Windichgratz, Rottemburg, Jörger, and Balbes.

Major Generals of Infantry: Mssrs. Dalberg, Marulli, Otto von Stahremberg, Langlet, and Diesbach.

2nd Line

Generals of Cavalry: Mssrs. Nadasdy and Merci

Generals of Infantry: Mssrs. Von Bevern and von Harrach.

Lieutenant Generals of Cavalry: Mssrs. von Lobkowitz, Frederick von Württemberg, Wachtendanck, Plichau, Gondrecourt, and Vétèrani.

Lieutenant Generals of Infantry: Mssrs. d'Aremberg and von Holstein.

General Majors of Cavalry: Emanuel of Savoy[12], Arragoni, Vobeseck, Leinbrück, Locatelli, La Marck, and Hamilton.

Prince Eugene forgot nothing to secure his camp and his communications. He gave his orders for the construction of the lines of circumvallation and from 20 June work began. The garrison executed a terrible fire on the workers and on the camp, but as they were far away, this fire had little effect. The Imperials working on the lines of circumvallation were somewhat more discomforted by the fire than the others. What was worse was that the Imperials were not able to respond to the Turks with equal volleys, as they lacked their heavy artillery. Prince Eugene sent orders to Peterwardein, Temesvar, and Segedin that they should send forward the men and materials that they had prepared.

On the 22nd, all was ready for constructing a bridge over the Danube on the side of the Belgrade Citadel, and work began as quickly as possible. The garrisons saw this and sent three large floating mills down the river, which were anchored under the cannon of the fortress. They caused no damage because little work had been done. The Imperials had the good fortune in stopping the mills and the Imperial saiques pulled them to the shore. However, this success did not stop the cannon of the citadel and the Turkish saiques from causing much disorder among the workers who worked on the bridge, and among the troops that supported them.

As the bridge over the Danube was constructed, the Imperials constructed yet another over the Save, which was to be defended by a redoubt manned by several troops and some cannon. The other bridge was begun in the swamps near the Danube to maintain communications with the region situated beyond the river.

[12]The son of Count de Soissons and nephew of Prince Eugene.

Prince Eugene still had no news on the approach of the Ottoman Army. All that his spies could tell him was that they had left their camp near Adrianople, but none knew the route they had taken. It was learned, for certain, that 13 Turkish ships had arrived at Semendria (Smederevo), and that they would soon be followed by a large number of others.

At the end of June, the Turks began to fire with their heavy cannon at the Imperial Headquarters and two shots fell near Prince Eugene's tent. They caused no damage, as they had lost all of their force.

While the bridges were being constructed a terrible storm arose which broke both of them. Several pontoons were detached and carried away by the violent wind.

The Turks seeing that the Imperials' communications were broken between the principal army and the Semlin camp hastened to profit from this opportunity to ruin the bridge over the Save. To this end they crossed the river in boats and landed 1,000 infantry and 150 cavalry, which then furiously attacked. The bridge was only guarded by 60 men from Hesse, who had only arrived at the camp. The Hessians, however, did not fail on this occasion to once again burnish their reputation for bravery, holding the position for a long time. They defended themselves with such courage and stubbornness, that Prince Eugene, who watched from the other side of the river, had time to send over more troops to assist the Hessians. The Hessian officer, who had defended the redoubt, received great praise and it was principally due to his actions that the bridge was saved. This was not a little service, for if the Turks had captured the redoubt, they would have been able to destroy the rest of the bridge and seize the pontoons.

Prince Eugene gathered up all of the forage on the far bank of the Danube to a range of six or seven leagues, so as to deny it to the Turkish army, which was said to be moving in that direction.

The Imperials continued to work on the lines of circumvallation and contravallation. The work went slowly because of the lack of fascines. The Turks had cut down all of the woods before the campaign and the Imperials were obliged to travel sometimes four or five leagues to find wood. Prince Eugene anticipated this and had ordered a number of troops and wagons to cut and carry to the camp a sufficient quantity of wood to the camp so that the workers found it ready for use and so that the troops would not be delayed.

While work on the lines was pushed forward the Imperials also worked on repairing the bridges that were damaged by the storm. The Save bridge was repaired at the end of June and a fort was constructed to defend it against the Turks and their ships.

The bridge over the Danube was also returned to a good state at the beginning of July. Communications were re-established and to better assure them a considerable reinforcement was sent to Count von Hauben, who camped near Semlin (Zemun).

Finally, the Imperials only awaited the arrival of their heavy artillery to begin opening the trench and word came that this artillery advanced as quickly as possible. Part of it was already at Titul.

The sight of the arrival of Count von Hauben at the Semlin camp greatly disconcerted the Turkish troops posted in that village and they abandoned the town on their approach. There were many skirmishes between the Imperials and the Turks, the Imperials winning almost all of them. The ships of the two sides frequently came to blows with much fury. My plan is not to enter into any detail of these small skirmishes. I will content myself with speaking of the preludes to a general and decisive battle, which I shall soon have the opportunity to address. Now, however, I will resume my account of the siege.

The Muslims had freshly raised, on the side of the citadel, three works filled with 50 cannon. They had a good entrenchment and a ditch. The Turks worked on other fortifications on the side of Belgrade away from the river, persuading themselves that the Imperials would not attack anywhere else. However, they were deceiving themselves. Prince Eugene had other plans. He was not unaware that the fortress was weaker on the water side, and it was there that he had resolved to attack. He made all of the necessary preparations. He began by attacking a fort that the Turks had raised across the Danube at the mouth of the Donawitz. Count von Mercy was charged with this attack. However, when the troops were moving to their attack positions, Mercy was struck with apoplexy. He fell from his horse without movement or consciousness. It was necessary to carry him off and delay the attack for a day. The detachment took up position so it could press the enemy, who on their side began a heavy fire and greatly discommoded the Imperials. It was initially thought that Count von Mercy was dead, but the strength of the medications given to him revived him and he later completely recovered.

Prince Eugene was informed of this incident and personally went to the attack site. He was nearly struck by three cannon shots that passed between him and Prince de Dombes, the terrible sound of which terrified their horses. Escaping this danger, Prince Eugene ordered the attack on the fort and after a short resistance it was captured.

On 15 July, the Imperials captured a new redoubt between Semlin and the island that the Turks occupied. Eight cannons were placed in this redoubt to keep the Turkish boats away. The warship *St. Eugene* was brought close to the redoubt as well. That same day a captain named Todor, returning to camp after a raid, related that in this raid he had encountered and defeated a Turkish party. He brought in 11 prisoners with him, and from them the Imperials learned that the Grand Vizier, at the head of the Turkish Army, had encamped at Nizza, where they had remained a few days; that a few days later they had departed and that the general rumor was that they had been given orders to relieve Belgrade, no matter what the cost.

On this word, Prince Eugene had his lines of circumvallation perfected and raised batteries. A prodigious quantity of fascines, gabions, and palisades were used in this effort. They had been prepared at Esseck and brought by water to the Imperial camp.

It was resolved, in a council of war, that the Imperials would take up a post beyond the Sava and establish themselves near its mouth. They intended to place the Bavarian troops there, which had only arrived at the camp, and that they would be reinforced by four regiments of cavalry: Darmstadt, de Martigni, Lobkowitz, and Savoy.

General Marquis de Marsilli was in command, with three battalions, six companies of grenadiers, and 1,200 pioneers with 300 masters (heavy cavalrymen), to work on the lines on that side, in order to cover the troops who would move there. Marsilli left with his detachment in the middle of the night and arrived in that position without the Turks perceiving it. The workers began to work without the least obstacle, until daybreak, when the Turks discovered what had happened. They then directed heavy fire from their saiques on this detachment, which was at the same time fired upon by the cannon from the fortress. However, this did not slow down the workers. They had already pushed a line and prepared rising some redoubts, when the Turks realized that if the Imperials established themselves in this post, their batteries could greatly inconvenience the Turkish ships, commanded 4,000 of their Janissaries to cross the Save. That they crossed the Save without losing a single man raised their courage. They marched with much resolution to cries of *Allah! Allah!*

The Marquis de Marsilli was not concerned by their cries or their proud countenance but realizing that his forces were not sufficient to resist an attack by 4,000 elite men from the garrison, he immediately sent a request for reinforcements to Count Rudolph von Heister, who commanded a corps of 300 men near him. Heister, however, did not want to come outside his defensive line of chevaux-de-frise, and informed Marsilli that he had to defend himself as best he could. On this response, the Marquis de Marsilli informed his men that their safety depended on their courage. He quickly arranged them to receive the Turkish attack. The Turks arrived, fired their volley, and drew their sabers. Howling terrible cries, they rushed on the Imperials, pushed them back, and put them in disorder. Marsilli was killed rallying them. His death increased their disorder, and they were about to break when Baron von Blumberg[13], Lieutenant Colonel of the Darmstadt Regiment, came to their relief with two squadrons. He charged the Turks and pushed them back. The Turks, recovering from their surprise, enveloped Blumberg and surely would have torn him to pieces (he was not supported by the infantry, which had broken and fled) if by good fortune Prince Eugene had not arrived at that moment. He came, according to his custom, to visit the posts and the works done during the night, and he had not expected to find

---

[13] Blumberg was promoted to colonel a few days later and his brevet was expressly declared to be compensation for his action on this day.

this situation. However, it was easy for a commander, adored by his troops, to repair such situations. The presence of the Prince stopped the fugitives. They reassembled and rushed back into the battle. Blumberg valiantly withstood the Turks, but he had need of a prompt assistance. That which Prince Eugene brought with him was sufficient. The Imperial infantry charged the Janissaries in the flank, broke them, and forced them to rush to the riverbank where they threw themselves into their saiques.

Prince Eugene did not amuse himself with pursuing them. He was content with having driven them off and quickly set about repairing the disorder that they had caused. He had chevaux-de-frise constructed, along with some entrenchments to cover the troops against any new attack, supposing that the Turks might wish to avenge themselves for this failure.

The losses suffered by the Imperials in this attack were considerable. Five hundred of their bravest soldiers were laid out on the square as were several officers of distinction. Count Rudolph von Heister, who had not wished to come to the support of Marsilli, was killed by a cannon shot which struck him behind his chevaux-de-frise. Count de Torrez, a Spaniard by birth, died as did Lieutenant Colonel Visconti and Baron von Sieger, Major of the Alt-Stahremberg Regiment. There were six captains and around 20 subalterns killed as well.

The loss of the Turks was no less. They lost the Pasha of Romelia, who commanded their detachment and who was esteemed as one of their bravest officers and the best known in the Ottoman Army. The Muslims were to regret the loss of so many men and the Christians mourned the loss of the Marquis de Marsilli. This action occurred on 17 July.

The Turks being repulsed, the Imperials completed their works beyond the Save. The trench was opened in this area and communication was assured by lines that ran from the bridge's redoubt to the trenches and beyond the Semlin camp. However, in order to not be exposed to sorties from the garrison, Prince Eugene reinforced the trench guard to nine battalions and three companies of grenadiers. He ordered that these troops remain all the night of the battle outside of the trenches in order to avoid any surprises. The Imperials now worked to construct a second bridge over the Save. Communications on the Danube were now assured, and a redoubt was raised on the edge of the river, to batter to pieces the fort that the Turks occupied on the neighboring island, which favored the retreat of the saiques of their fleet. Twenty-five cannons were placed in batteries facing this fort and they did not cease to fire until the Turkish fort was battered to ruins and the Turks had abandoned it.

On 22 July, all of the batteries that were aimed to fire on the city were in a state to begin fire. Nearly 30 cannon and 15 mortars began to fire on the 23rd. The cannon fired on the citadel as well as the front and reverse of the principal defenses. As night approached bombs replaced balls and spread desolation and death into the city. The ravages in the city were terrible as the streets were very narrow and the houses badly constructed. The ruin of one house would provoke the collapse of its neighbor and three or four houses would be destroyed by every shot. It was something terrible to hear the cries of the unfortunate people who did not know where to flee. Some were crushed in the ruins of their houses, while others went through the streets seeking asylum and finding death under the walls of other houses that collapsed on their heads or by stone fragments that filled the air. The garrison responded with a furious fire, firing many bombs, stones, and shot against the Imperials, but with little success. The bombardment lasted two days. The Imperial cannon dismounted many of the garrison's batteries, whose men had the chagrin of seeing themselves destroyed without causing the least damage to the Imperials. Soon Belgrade no longer resembled, on the side of the river, a collection of odd hovels that time had greatly abused. However, on the inland side it was different. Not only were the fortifications still in a good state, but the Turks continued working on them as the Imperials were completely focused on the waterfront, appearing not to care what went on elsewhere.

The fracas that the bombs had made in the lower city appeared to have broken the spirit of the garrison. It made no more sorties, and it might even have been thinking of capitulating if the major part of the outer defenses were not still in a good state and they had not received certain news of the departure of the

Ottoman Army from Adrianople and the extreme diligence they were making to deliver the fortress. Be that as it may, Prince Eugene also learned that this army had arrived in the vicinity of Nissa. Some say it contained 250,000 men, while others claim it contained 300,000.

About this time there was a rough fight between the Turkish ships and those of the Emperor, commanded by Admiral Anderson. The Turks were beaten. One of their saiques was sunk and several others were so badly pounded that they had to be towed into port.

Baron von Petrasch, who had undertaken a skirmish war with his hussars, wrote to Prince Eugene that he had advanced to Sabatz at the head of 300 horse, resolved to capture the post. However, when he got there he discovered that it was well provided with cannon and a large garrison. Petrasch, having neither infantry or artillery, judged it proper to abandon his plan and posted himself between Sabatz and Mitrowitz in order to observe the Turks and to favor the communications between the Imperial Army and Peterwardein. He added that the Muslims had published rigorous orders throughout Bosnia obliging the inhabitants who were in a condition to bear arms, to take them up and send all of the healthy young men to join the Ottoman Army.

Prince Eugene received other intelligence, which informed him that this army had decamped from Nissa and after having crossed the Morave, they had advanced to the stockade of Haffon-Pasha, a small post about six leagues from the Imperial Camp. Prince Eugene placed cannons in his lines of circumvallation.

On the 28th, he learned that the Turkish Army was marching on Semendria and that the Janissaries had already arrived at Crutscha. Eugene redoubled his precautions for the defense and protection of his lines, in the fear that the Turks would attack them with cold steel, which was their ordinary manner.

The next day a large body of Turkish horsemen advanced to within half cannon range of the Imperial camp to examine the entrenchments. A few shots were fired at them to encourage them to move away. Prince Eugene had powder and shot distributed to the troops and all dispositions necessary to receive the Turks were made. At the same time, he sent a courier to Vienna to inform the city of the Turkish Army's approach and that one order the prayers of the entire Empire for the success of the Imperial Arms.

On the 30th the vanguard of the Ottoman Army appeared on the heights before Belgrade. The Pasha, who commanded the fortress, expressed his joy with discharges of the artillery on the ramparts of the city and by a sortie, which was repulsed with significant losses. The volunteers in the Imperial Army, and those of the Turkish Army skirmished and fired their pistols at each other throughout the day. Ibrahim Pasha, one of the bravest Turkish officers was killed in one of those skirmishes.

The next day the Imperials watched the Turkish Army arrive in large bodies on the heights and pitch their tents. On 1 August, they appeared behind their flags on the Crutsches Mountains, which formed an amphitheater which presented the most beautiful and terrible view one has ever seen. This army was not as numerous as had been reported. An aga taken by an Imperial detachment unwittingly provided an accurate count. In one of his pockets the Imperials found a state of the army, which Prince Eugene had translated. It consisted of 80,000 Janissaries, 10,000 Asian militia, 10,000 European militia, 30,000 Tartars, and 20,000 Spahis, most of which were new levies.

Although this letter reduced to 150,000 men that which had initially been reported, it was still a formidable force. Prince Eugene was enclosed on the right and left by the two great rivers; in the front by a city which had a garrison of nearly 30,000 men, and to the rear by an army of 150,000. It is true that he had bridges over the Save and the Danube, but that was not a great assistance as if he made the least move to the rear, he risked being charged, seeing his bridges broken and burned, his troops drowned or their throats cut by the Muslims. His army was already greatly reduced, be it by the fatigues of the campaign, or by disease, or by battle with the Turks, or by detachments that he had been obliged to make to reinforce the various corps destined to cover the region. His only task now was to remain behind his defensive lines until the city surrendered. This could not be long, even though the Imperial troops that attacked from the water side were greatly fatigued. This part was subjected to the most terrible disadvantages. Remaining in the lines situated in a valley dominated on one side by the ramparts of the besieged city and on the other by the little hills where the Turkish relief army was encamped it was also subjected to the fire of 300 cannon and more than 80 mortars, without being able to respond. Nonetheless, Prince Eugene took this last part, as the worthiest of his glory, because he had come to take Belgrade, and when he could have retired without difficulty, perhaps he did not wish to do so, persuaded that such an act would damage his reputation, which was founded on a series of victories. He had little difficulty saying to his generals that he would take Belgrade, or the Turks would take him, wishing to testify thus that he would sooner lose his life and his liberty than stop the enterprise on which his glory depended.

All of Europe, informed of Eugene's situation, trembled to see him succumb. He found men who accused him of boldness, of having undertaken a siege that appeared impossible. Others accused him of presumption, of allowing himself to be enclosed by such a numerous army that could destroy him with their artillery alone. Finally, everyone reasoned in his way and according to the passion that dominated him, or according to the passion of those who wrote from the Imperial camp, thus the people who were distant from the scene received different impressions. Despite all of this, one could still find an infinity of people who recalled the great actions of the Christian heroes and had no doubt he would defeat the Turks. These different ideas gave rise to many disputes and challenges in all of the cities of Europe. Some contended that Prince Eugene would take Belgrade, while others said he would not. The Imperial Court, as the most interested, was also very alarmed, because ordinarily, they had more interest in the situation than fear of it. The Emperor wavered between fear and hope. On one hand, they were aware of the bad state of his troops; but on the other they knew Prince Eugene's talents, his genius so fecund in resources, his prudence, and his valor, which gave some rays of hope. Among the generals of the army there were two very opposite sentiments. Some, alarmed by the terrors that the envious of Eugene spread throughout all of the camp, absolutely despaired ever coming out of that bad situation. The others, friends of the Prince, or those who were attached, said boldly that their General knew well how to pull them out of this situation. And when others objected that the Turks were reinforcing themselves every day and their numbers grew as they watched them: "So much the better," they said, "the more of them that there are, the more we will kill." The simple soldiers had no such hopes. However great was the danger, they did not doubt that Prince Eugene would find a way out of the situation. Those who had served under this hero, shared with the young stories of the great battles he had fought, and they burned with impatience to see how his project would turn out so they could be witnesses to his glory and to share the danger with him.

However, these flattering ideas suddenly changed to despair. From the end of July Prince Eugene fell sick with a fever, which obliged him to take to his bed. All the army was in a strange consternation. One saw terror painted on the faces of officers and soldiers. "What will become of us if he dies?" they asked of one another. "Who will lead us against the enemy, who will take us out of this place?" "The Muslims will not wait for that time to fall on us and do so with much audacity, that they will not fear having a battle with a general whose name is so terrible to them."

The Viennese Court, which during the Prince's illness, did not receive any reassuring news from Hungary, was not in the least alarmed. It learned that the Prince was sick, but they did not know how great his sickness was. And ordinarily their fears grew as everyone imagined the worse. Charles VII withstood some uncertainty with his sentiments of piety and religion, which Europe still admires in him. Entirely resigned to the will of God, he impatiently awaited the denouement of such incidents. However, he forgot nothing to draw the favors of Heaven on his arms and on the days of the general, who commanded them. One saw edicts from the Emperor calling for processions and public prayers and above all of the reforming of customs. He gave personal examples of the greatest devotion by his numerous acts of piety and by the humility with which he appeared at places consecrated to Christ.

The Prince's illness passed and there were no serious repercussions. On 2 August he wrote the Emperor, "The Turks, have come to examine us, and on that they have chosen a camp which they occupied in the evening, which is directly posted in the front of our entrenchments, extending from our left wing by a valley to the height of our right wing.... We still do not know what they plan to do with another corps which they have detached towards Mehadia. Deserters tell us that the Grand Vizier has received orders from the Grand Seigneur to attempt at whatever price, to relieve the fortress. It appears that the lack of forage obliged them to this step and that without it they could not long remain here. For us, we have made all of the necessary dispositions to receive them warmly and to hold ourselves ready for all events, if they move to attack us.

One sees in this letter that the Turks had sent a detachment to Mehadia, a small post, or type of stockade five leagues from Orsova. It was, to some degree, the key to the Temesvar Banat, and the Turks took it with the objective of ending raids into that region, to ravage it, and then take their predations beyond the Aluta into Wallachia and Transylvania. They had no difficulty in making themselves master of this fortress. It is natural that 20,000 men can overcome a garrison of 500 men. That of Medahia had no more than 500 men and the corps that attacked it was at least 20,000 men.

The Grand Vizier received, finally, the heavy artillery that he had been awaiting. One saw the Turks soon entrench themselves on the heights and construct batteries for their cannon and mortars. Prince Eugene arranged his own artillery to reduce the upper city to the same state as the lower city. He had created a sufficient breach to launch an assault, but he did not have the forces necessary to do that in the presence of a relief army. It was necessary to reduce Belgrade with bombs and cannon and starvation. Hunger was already beginning to be felt. The Imperials, masters of the river, occupied all of the avenues on the shore and had blocked the movement of supplies to the garrison and it was easy to see that they were being pressed, since one saw from time-to-time rockets coming out of the citadel from time-to-time, which were signals to the Grand Vizier to hasten the delivery of the fortress. However, the Grand Vizier did not push his forces. Pleased with the advantageous situation of his camp, he only wished to be able to subsist there for a long time so as to give his artillery the leisure to exterminate the Imperials.

It was on 3 August that the Turkish artillery began to fire on the Imperial camp. Two batteries raised on the two heights saluted the Imperials in such a furious manner that the Imperials recognized how much damage the other guns would cause when the Turks had all of the other guns operational. The shot that fell from high to low knocked down tents, men, horses, and in a word, whatever they encountered. The bombs which fell everywhere struck what the cannon could not. Nobody knew how to put themselves under cover from the hail of shot. No matter where one turned, one found death. If one fled to the side of the city, one was exposed to fire from the city's guns. In the camp, there was no asylum against the Turkish Army's cannon. The King's quarters were also exposed to their brutality, which required it to withdraw. Prince Eugene's quarters suffered no less than the others. Several of his servants were killed near his tent. The Prince did not wish to change the position of his quarters, and it was only after the repeated requests of his generals that he agreed to have his quarters moved to a less exposed position.

The Prince had his trenches raised to the measure that the Turks augmented their batteries. He distributed gabions to the troops to cover them as much as possible, but all of this did not stop the Turkish guns from doing more damage and it was impossible to avoid. One saw Imperial troops that had, at the beginning of the campaign, formed a beautiful army, decline daily. The Turkish cannon carried away entire ranks and bombs killed or maimed many men. This went on for four weeks and each day the Imperials buried hundreds of men. Contagious disease began to appear among the horses which died in numbers every day such that in less than three weeks part of the cavalry was dismounted.

Prince Eugene saw all of this with a sorrow that was as easy to see as it was to express. It required everyone to stop this. The steps that he took to this end could not be too great. He anticipated the needs of everyone. Remedies were given by his order. Bread, meat, and everything that was necessary for life was in abundance in the camp. The soldiers, observing the attention of their generals to protect their lives, supported the situation with patience, and died with the only sorrow that they had not died in battle. Everyone judged that if the Imperial Army continued before Belgrade, it would be entirely ruined and perhaps not in a state to defend its entrenchments.

To that point it was expected that the Grand Vizier would be obliged to decamp, because of a lack of food and forage. With regards to this last point, it was known that the Turks had to be in great distress over the lack of provisions since the Imperials had stripped the region for seven or eight leagues around Belgrade, and since there were nearly 50,000 horses and camels in the Turkish camp, so it would not be easy to keep them fed.

Nonetheless, the Turks did not appear much embarrassed on this article. It is an advantage that this nation has over all of those that surpass it in the wart of war. It can feed its cavalry where those of the Christians would die of hunger. A bit of chopped straw, with a little bit of grain suffices to feed the Turk's horses. They could feed 100,000 horses on what 50,000 German horses or 25,000 French horses would infallibly die, lacking sufficient suitable food.

As the Imperials observed the consistency of the Turks as they remained at their posts and their desire to raise the siege, they expected them to come down from the heights and fall on them at any time. Even so, Prince Eugene doubted it, when he considered the strength of his entrenchments, which the Turks had had the leisure to examine. Nonetheless he distributed ammunition to the troops destined to the defense of the lines and the cannon were loaded and ready to thunder on the Turks. The Turks forgot nothing to strike their blows without risking anything. Their batteries continued firing on the Imperials and they had pushed their lines to a hill near the Save. Prince Eugene, always watching the Turkish movements, understood the ramifications of this. He understood well that by means of this hill, the Turks could easily ruin his bridges over the Save, or even pass, without losing many casualties, 30,000 men over the river to strike the corps of Count von Hauben, commanding near Semlin, who would be destroyed before he could be assisted, or they could attack the other troops that were destined for the attacks on the lower city, which were separated from the main army. If the first of these things occurred, the Imperial Army would find itself reduced to the most desperate state that one could imagine in the case of a forced retreat. And if it was the second, then the army could be beaten in detail and the campaign lost. Thus, the war, which was initially favorable for the Christians, would become fatal for them. The Venetians could not withstand the Turks. The Emperor and the Empire were already exhausted in their war with France that had just finished, so they were not able to provide for the immense expenses of this one. This reflection and many others of this nature caused Prince Eugene to know, that to draw all of these powers and himself from their embarrassment, it was necessary to provoke a decisive battle. However, as the fate of the Empire was at risk, it was necessary that they take the steps necessary to assure success. Eugene resolved to wait a few more days, hoping that the Turks would launch some action or make a movement from which he could profit.

However, the Turks had opened a trench, according to their custom, before Eugene's camp, and were slowly approaching by means of deep trenches, advancing their batteries to the measure that they pushed their works forward, such that soon they would be at a point where they could fire their cannon at the range of a musket against the Imperial lines, producing crossfires and, as a result, causing much more damage than they had been able to inflict before.

The Imperials, on their part, continued to fire on the city as well as the Turkish Army, but with more success on the former. The reason was clear. The Imperial camp was on the same level as the city, but to fire on the Turkish Army, it was necessary to fire upwards. A bomb thrown from one of the Imperial batteries fell on a magazine in the lower city and completed the destruction of the few houses that remained standing in that quarter. The magazine exploded with a terrible blast and 3,000 people were buried under its ruins or under those of the nearby houses as they collapsed.

The Turkish army was shocked by this and ran to arms without knowing what had happened and appeared to arrange itself in battle all along its front parallel to its trenches and crying aloud. Prince Eugene had so well disposed all things in his camp that in an instant his troops were ready to receive the Turks. However, the alarm passed. The Turks attempted nothing significant. They contented themselves with firing their muskets on the Imperials, who responded with much vivacity. This fire lasted for most of a half hour. The Turks then re-entered their camp, leaving only a few cavalry to skirmish with the volunteers of the Imperial Army, who had come out in great numbers, supported by a part of hussars, ready to charge the Turks.

Prince de Combes exposed himself to an extreme degree on this occasion. He and his suite were long exposed to a battery of cannon that broke the thigh of Count d'Estrade, his governor, and the foot of a page on whose shoulder the young Prince was supporting himself.

The Turks constantly pushed their batteries forward; Prince Eugene's quarters found themselves exposed to their shot, which fell on his tent more than ever. All of the generals ran to beg him to permit that his quarters be moved to the left wing, where the Turkish shot could not reach. The Prince could not resist their requests and allowed it.

The Turks noted these changes and sensing well the greatest blow that they could strike would be to kill Prince Eugene, now directed their efforts to the left wing. They gathered up a great number of fascines to fill the ditches of the Imperials' trenches and the swamp that covered them on that flank and worked to move their trenches closer on that side, as they had earlier done on the right.

Around this time Prince Emanuel of Savoy was brought back to the camp by some soldiers who had found him in a region where his horse had thrown him. His injuries were such that he was transported to Peterwardein spitting blood and suffering from a fever.

Before entering into the details of the dispositions that Prince Eugene would make for his attack on the Turks and before giving an account of the famous battle of Belgrade, I am persuaded that one will not be unhappy to find here a letter which appeared in the *Nouvelles Publiques* of that time and which was written by a French volunteer serving in the Imperial Army. The style is not particularly flowery, but in compensation one can see that the author knew well his profession.

"Finally, the Ottoman Army," he said, "which had threatened us for a long time, began to appear on the 28th of the last month and grew in numbers over four days; it moved to camp on the front of the flag, during the 31st, on the heights within range of our cannon. We saw then a camp filled with beautiful troops and guards on their flanks which appeared prodigious to us, without, all of the same, being ever able to discover their true strength and numbers from any deserters or prisoners. We judged only that this army could be about 200,000 men. We thought initially that a large number of troops could not be sustained eight days in this camp, having neither water nor forage; but experience told us the contrary.

"The Turks began to dig in the same night they encamped and advanced trenches and parallels towards us. They raised, in a very short period, mortar and cannon batteries halfway between us and each day they fortified themselves with some new works. The front of their parallel occupied clear ground large enough for 20,000 men in battle order and they placed, in this interval, 140 cannon and 35 mortars, which fired upon us, from the 3rd of that month, almost along the length of our front. This obliged a great part of our men to rapidly withdraw and seek cover in the traverses, the parapets of our trenches, and the King's quarters found itself obliged to evacuate. After this movement, the Turks continued working under the fire of their cannon, approaching our ditch by an infinity of branches, which, though badly organized, nonetheless allowed them to fire a great deal. It was then that these works, the parallels, were joined after a while, causing us to see our army as regularly besieged as one besieged a fortress; and I can assure you that a spectacle so new did not fail to give some concern to the veteran officers on the outcome of such a great affair as that for which they prepared.

"Never, in effect, at least as far as I know, had anyone seen[14] an army besieged in its camp, while it besieged a place as important as Belgrade. It is, however, certain that the Imperial

---

[14]This officer apparently had never served against the Turks before this campaign; because that he would have had occasion to see this method of fighting on the last occasion at the battle of Peterwardein. Translator: Both men are apparently unaware of the battle and siege of Alesia, fought by Julius Caesar against the Celtic leader Vercingetorix in 52 B.C.

Army, which besieged this city, was for nearly 15 days besieged in its own camp, by an army of at least 150,000 Turks, and supported by a numerous force of artillery, caused our troops a considerable loss. It is true that our general fired against the enemy with 85 guns, both cannon and mortars, which were placed in various batteries along the line; but as they were firing from low to high, it is easy to judge that we did not cause as much damage to the Turks as they caused to us. The situation where we found ourselves for

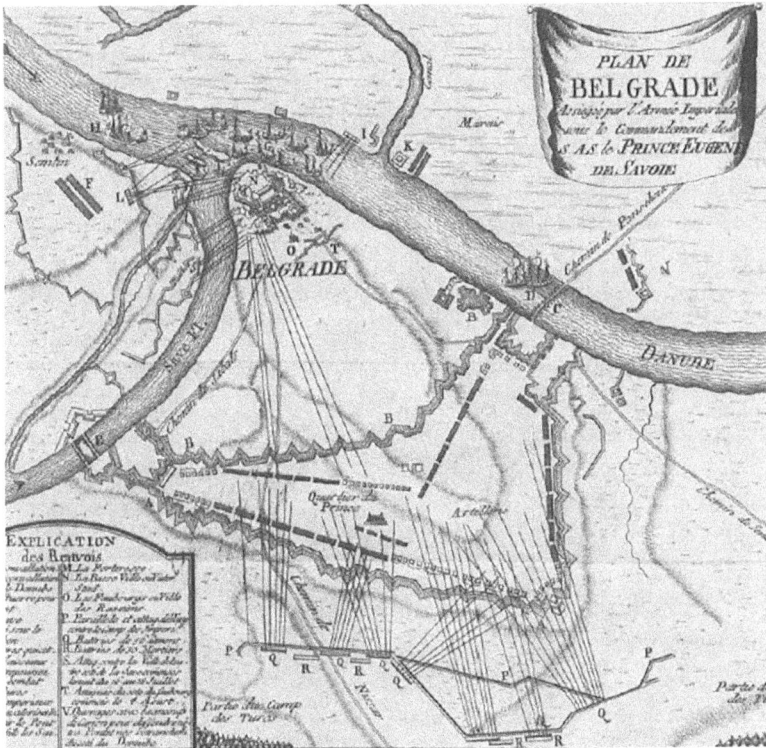

Attack on Belgrade

a long-time produced fear in our men that the Turks would force some point in the trenches that covered the camp and that our army was in danger of dying; because its only line of retreat was over the Save which could quickly become useless. However, the confidence of all of the army was justified by the skill of Prince Eugene, and in the force of our entrenchments, which the enemy would be obliged to climb, on which we had taken just precautions, to always assure them. In addition, our troops were convinced, up to this time, that the lack of water and forage, which the enemy were obliged to seek at a distance, would not long permit them to remain in their camp and it is this which had constantly held tranquil.

"However, in such extremity as were the Turks, it is certain that before yesterday (13 August) the fire of their musketry fire already was passing far beyond our parapets[15] and they were within range of seeing that we had made no movement to interrupt them, so their audacity was greatly augmented, that in two days they had pushed new parallels within pistol range of our entrenchments, and they had removed, at least according to prisoners, into the ditch on 17 August, and that they were making major preparations. It was added that they were ready to roll large gabions before then, which would put them under cover, and that they would support this with fire from their parallels, and the same gabions would then serve to fill in the ditch throughout all of its length and over all of the width of their attack.

"But the question is to know if our general will give them the time to execute their plan. It is this which is very doubtful, because Prince Eugene, seeing himself so closely pressed, will find it necessary to come out of his trenches, despite the resolution that he had taken to maintain himself and let the enemy attack him; and it is for this reason that he held a grand council of war, where it was decided that it was necessary to go against them, and fight them tomorrow at daybreak. He also moved to make public this enterprise and he had resolved that we would have a first line, composed of 30 battalions and 24 cavalry regiments, each of six squadrons, and divided on the right and on the left of the infantry; that this first line should be supported by a second of 27 battalions; all of this being under the orders of Marshal von Palfy, Prince Alexander von Württemberg, and Count de Merci; and that the rest of the troops destined for the line of circumvallation, should line the parapet of the entrenchments in case some unfortunate circumstance required the army to retreat.

---

[15] Translator: A musket had a range of no more than 100 paces, so the Turkish trenches were *very* close to the Imperial trenches.

"Finally, he had resolved that one would begin to defile tomorrow at 2:00 a.m., through the different barriers, to be able to form before the enemy before daylight disclosed our movement and plan, and that the signal for the attack on the Turks, in taking the right and left flanks of their works, would be the sound of three bombs which would be fired simultaneously from our mortars. The great object of the generals was, therefore, to sweep, if they could, all of the trenches and penetrate to the enemy batteries; and if they could reach there, to form there in a good order for battle, and to cover our troops until they had filled in all of their trenches; because they counted on doing much, if with 35,000 effectives, which is all that we had for this sortie, we could succeed. This affair promised to be very serious. We had 80,000 Janissaries to fight and posted behind trenches. However, we hoped everything would go well, anticipating that the besieged [garrison] would not make a sortie to discommode our projects. It is true that we left in battle some troops that were destined for the lines of circumvallation, to oppose any movements that might come from the city while we were in battle."

After the Imperial generals left the council of war, where it had been resolved to give battle, they went to work to execute the Prince's orders. All the detachments that had been dispatched to various locations were recalled. One thousand infantry and 300 horse alone were left on the far side of the Save. The Imperials left 300 infantry on the island, now called the "Bohemian Island," and 1,000 in the redoubt on the shore. Seven regiments of cavalry and eight battalions, with four companies of grenadiers were designated to guard the lines of circumvallation. These troops were commanded as follows: the cavalry by Lieutenant Field Marshal Baron von Viard, who had rejoined the army a few days earlier. He had requested of Prince Eugene the right to fight, but the Prince had responded to him that if someone had to oppose the besieged city's garrison, who would be better than he. Viard had under his orders Generals Lanthieri and Orsetti. The infantry was commanded by Lieutenant Field Marshal Count von Brown with General of Battle Wobese. Two battalions were left at the bakery, and one gave orders to the dismounted horsemen and dragoons to take positions in the lines of circumvallation and contravallation.

The rest of the troops were destined to form the lines that were to march against the Turks. The cavalry on the wings was under the command of Field Marshal Count von Palfy and the infantry in the center was under the orders of Field Marshal Prince Alexander von Württemberg. The right wing was formed of 11 regiments of cavalry and dragoons.

The first line of this wing was to be commanded by General of Cavalry d'Ebergeni, with Lieutenant Field Marshals Count von Hauben, Lobkowitz, Prince Frederick of Württemberg, and Generals of Battle Galbes Jörger, Uffeln, and Arroi.

The generals of the second line were Counts de Merci, de Croix, Hamilton, von Wehlen, la Marche, and Elz.

The left wing was to contain 12 regiments of cavalry and dragoons, and the first line was to be led by General of Cavalry Count de Montecuccoli, who had under him Lieutenant Field Marshals Walmerode and Hautois, with Generals of Battle Cordoua, Rottenahn, Arrigoni, and Windschgratz.

The second line of the left wing was to be led by General Count de Martigni, with Lieutenant Field Marshals Veterani and Gondrecourt, and Generals of Battle Eck, Locatelli, and Zollern.

The battle corps, or the center of the army, consisted of 22 battalions and 23 companies of grenadiers.

The first line was commanded by General of Artillery Count von Harrach, with Lieutenant Field Marshals Count von Daun, Maffey, and Bonneval; and Generals of Battle Dalberg and Merci (the Elder). These were the generals of the left wing of this corps. Those of the right wing were Count Maximilian von Stahremberg, General of Artillery; Lieutenant Field Marshals Wachtendank and the Duke d'Aremberg; and Generals of Battle Langle and Leimbrück.

The second line has its commander General of Artillery Prince von Bevern, Lieutenant Field Marshals Duke of Holstein, Wallis (the elder), and Pischau; and Generals of Battle Merci (the younger), Ottokar von Stahremberg, and Wallis (the younger).

The Imperials formed a reserve corps consisting of nine battalions and eight companies of grenadiers, which was destined to remain in the lines with orders to hold itself ready for any event. Lieutenant Field Marshal Baron von Seckendorff, who had greatly distinguished himself during the siege, commanded this corps. He had under him Generals of Battle Count von Diesbach and Marulli.

All of these troops came to a total of around 60,000 men, but the two lines were designated to act alone in the battle, leaving more like 40,000 to actually do the fighting.

The two sides were far from equal in number. The Turkish cavalry alone was nearly as strong as the Imperial Army. But despite this, there was no one in the Imperial Army, from the lowest soldier to the general officers, who did not hear the decision for battle from the council of war and see the dispositions for it with joy.

The soldiers who had watched their comrades killed by Turkish artillery far preferred to die with their weapons in hand. "At least," they said to their fellows, "we will make them pay with blood." One only had to examine the situation of the Imperial Army and the evils to which it was exposed to judge the desire to come to battle in each soldier. In effect, what could be more natural than to prefer a doubtful and distant danger to a present and nearly certain death? What would be more reasonable than to look to die with glory on the battlefield, than the oblivion behind the lines where one had no means to defend oneself?

It was with these general ideas that all of the army was transported with joy when they heard on 15 April that there would be a battle the next day.

Never had one seen such good will. The soldiers flew to their assigned posts. The volunteers asked that they be allowed to form a separate corps, in order that their actions could not be confused with those of the others, but Prince Eugene refused them, so as to not expose the young men of the first rank, who found themselves in numbers among the volunteers. Prince Eugene ordered that the volunteers be mixed with the Imperial squadrons and kept by him only the Princes of Bavaria, those of the Blood of France, those of Lorraine, and the principal other young lords who the love of glory had brought to Hungary.

Part of the night was spent in arranging the army for battle. The order was that the first line would step out one hour after midnight; that the right be supported on the flèche of the entrenchments; that the left would extend towards the plain where it was to direct the main attack, and that the second line would follow to support the first, regulating its movement on the first line such that it retain, in marching and arriving at the Turks, the same distance from the first line that it had when it arranged itself in battle before moving on the Turks.

Prince Eugene did not imitate Alexander the Great, who slept deeply the night before the great battle of Gaugamela, also known as Arbela. Eugene knew too well that vigilance must be the principal quality of a general, and no one, as I have already commented, knew this more than he did. He was alert all night, moving from point to point, exhorting his officers to do well, recommending great silence to everyone, and freely giving to the soldiers everything that they required. Wine, beer, and cognac; nothing lacked to fortify the men, and put them in a state to support the fatigues of the day that was about to begin. Nothing, however, could be done for the horses, because of the lack of forage. As for ammunition, it was freely distributed, and the ships had brought it in such great quantities that the soldiers were well supplied. Powder and balls were given to whoever requested them.

The Imperials continued firing bombs into the fortress[16], during the night, to amuse the Turkish Army, but around midnight this bombardment stopped as the gunners recovered their breaths. Then, an hour later, everything was ready for the attack. The signal of three mortar bombs was given and the battle began.

The first line immediately put itself in movement with the least noise possible. It advanced slowly in the light of the moon, but a thick fog suddenly arose, and the right of the line strayed and, instead of supporting itself on the fléche of the entrenchments, as ordered, it encountered a Turkish trench.

It appears difficult to understand how the Imperial Army could make such great movements and how the generals could give so many orders without the Turks having the least awareness of the situation, but they had not the least suspicion of what was coming. However, this might be because the Turks made little use of spies, which they paid very badly, and because the Grand Vizier and all of the Turkish Army was so persuaded that the Imperial Army was reduced to less than half its original strength and that it suffered the last extremities, that they would only ask for quarter and were far from thinking of fighting. Whatever may be true, they committed the error of neglecting their security. The Turks had taken no steps to be informed of Prince Eugene's plans and actions, so they could attempt to check them. This was the first reasonof their defeat at Belgrade.

However, the Imperial right wing had fallen, as mentioned earlier, into a Turkish trench during the darkness produced by the fog. There was some initial confusion, but they held firm. Be it because they thought it some party or patrol from the Turkish

---

[16] Translator: This was probably also done to cover the noise of the troop movements as they prepared for battle.

camp or be it because it was a movement of courage that caused them to act, they ran to the arms that they had grounded and executed a volley on the Imperials. Count de Palfy, who was there, and who had his cavalry moving with their musketoons in hand, immediately ordered them to fire. They fired immediately. This was the last signal for the battle. On hearing this volley, the Turks began their terrible battle cries, which spread alarm throughout their ranks and was repeated by the echoes in the nearby mountains, resembling the waves on an agitated ocean, or an earthquake, which sometimes seemed about to submerge Sicily. What augmented the horror of the tumult was the thickness of the fog. One could not see 10 paces and this obscurity joined to the surprise of the Turks, and the little harmony that they observed in forming their battle formation, which would have formed a spectacular sight, had the fog permitted one to see it. One saw Turkish soldiers embracing each other, striking each other, and running here and there like mad men. Turkish officers shouted orders to soldiers who fear, or surprise had rendered deaf; one heard them screaming for their troops to follow their commands, without response. In a word, the Turks had fallen into complete disorder. And it is certain that if the Spahis and the Tartars who were already on their horses behind the lines and in the works, had not held firm this time, which was contrary to their custom, one might have seen an army of 150,000 flee after a few chance shots were fired by an army of 40,000 men. However, the Spahis and Tartars, having received the Imperial cavalry bravely, the Janissaries had time to rally. They were seen running forward, or better to say, they were heard rushing forward in a crowd in their trenches and attacking in large platoons the Imperial troops.

The battle escalated and muskets began firing on the left and in the battle corps. The cavalry of the Imperial right, which, as I have noted earlier, had gotten lost and pushed forward after having passed over the stomachs of the Spahis and Tartars, who had fought it only for a while. They were little inconvenienced by the ditches and trenches which they encountered, from which the Janissaries fired constantly, as did the field artillery, but that did not stop it from advancing.

The infantry of the right wing had followed the cavalry, supporting it bravely, and one could say that it saved the cavalry with its fire. However, the success that it achieved was well balanced by the disadvantages that it caused, because in pushing away from the path that had been designated for it, it opened a gap capable of holding several battalions.

Prince Eugene, who had not chosen a particular spot from which to observe the action was then at the extremity of the battle corps, surrounded by all of the volunteers of the first rank. The fog prevented them from seeing what was happening on the right center, and the gap remained. The Turks finally discovered it and threw themselves into it in great numbers. The outcome of the battle became doubtful. The Imperials, taken in the flank and rear were on the point of succumbing, when suddenly the fog vanished, and they discovered their danger. It was great good luck that the fog that had obscured the sky dissipated at precisely that moment. A Spanish historian called it a miracle and claimed it had been done by the Virgin Mary or by Saint James of Galilee or Saint John of Capistrano.

Prince Eugene, observing the disorder on the right of the battle corps, rushed forward the second line, and placed himself at the head of the troops, charging the Turks followed by the volunteers. The Turks thinking victory was theirs, refused to let it slip away. They held firm, and the battle became bloody. The Turks pushed fresh troops forward to support those that were fatigued. Prince Eugene, lightly wounded by a saber blow, redoubled his efforts. The German soldiers seeing him exposed in the middle of the melee threw themselves into the fray, each wishing to take part in it. Never had troops fought with such valor as the Germans in this battle.

The efforts made by the second line to push back the Turks in the center were such that after having made great carnage, it obliged them to return to their trenches and the gap was closed.

Prince Eugene, who had seen that the ardor of his soldiers had caused the disorder that he had repaired, he sent orders to the left that no one push a single brigade ahead of any others, and that everyone charge at the same time. However, despite the efforts of his generals to conform to this, it was not possible for them to hold back the Bavarian infantry. This brave infantry, carried away by a noble emulation, continued marching forward, despite the difficulties that it encountered. A French gentleman named "la Colonie," a general officer in the service of Bavaria, who commanded this infantry, far from holding them back, was the first to excite them. They crossed the ditches and ravines that were there in great number on that side, pushing over parapets and a thousand over obstacles behind which the Turks covered themselves. They engaged the Turks, charged them, and threw them back. The Turks fled trench to trench. They sought to conceal themselves and the Bavarians pursued them, cutting them down with their bayonets and sabers.

The Bavarians profited from their success and seeing themselves supported by some brigades and diverse cavalry regiments that had come to their support, marched on a battery of 18 cannon that was greatly discommoding them. It was not easy to capture the battery as 20,000 Janissaries and more than 4,000 Tartars guarded it. However, as all of the cavalry and infantry of the left wing had marched on it, following the path opened by the Bavarians, they were within range of supporting the Bavarians as they marched on the battery. Prince Alexander of Württemberg advanced the second line infantry to reinforce the attack of the Bavarians. The Turks, after a light resistance, abandoned the battery, which was immediately turned against them.

The same was done on the right wing, where Turkish batteries had also been captured. It was then that victory was suddenly secured. The Turks were driven back on all sides, pursued to the heights, and then into the plain. There they rallied in some force and their cavalry turned about to the right, moving to envelop the three regiments of German cavalry that had pushed too far forward. One immediately fell into disorder and part of the horsemen were cut to pieces, but the two others defended themselves so well that they gave time for some dragoon regiments to arrive and disengage them. The Turks now only thought of fleeing, leaving behind their camp and everything it contained. The Rasciens and the hussars, which were sent in pursuit of the Turks, executed great carnage on the fleeing Turks, granting no mercy, even to the wounded.

The loss of the Turks was great. It is thought they lost 10,000 dead, 5,000 wounded, and as many prisoners on the battlefield. It is estimated the 3,000 further were killed in the pursuit.[17]

Their camp was entirely deserted. "Except that," wrote a historian[18], "it resembled a great city and it was filled with an infinity of provisions and munitions. All the tents were new as were the wagons, the equipment, and artillery."

The Imperials lost 2,000 dead, plus 3,000 *hors de combat*, of which 1,800 recovered from their wounds. Many men of distinction were lost, including General Count von Hauben, General Dalberg, the young Count Palfy, Prince de la Tour-Taxis, the Marquis de Clerici, and Vilette, some colonels and some other senior officers. Among the wounded were Prince Eugene, Prince Frederick of Württemberg, Field Marshal Count de Palfy, Prince von Lobkowitz, Generals Erbegeni, Rottenhan, Locatelli, Arrigoni, and several other officers of distinction.

The booty that was found in the Turkish camp consisted of 131 bronze cannons, 30 mortars, including some that fired 200-pound bombs; 20,000 cannon balls, 3,000 bombs, 3,000 grenades, 600 barrels of powder, 52 flags, 9 horse tails, four trumpets, a great Janissary drum, a number of smaller drums, a great Spahis kettle drum, another smaller kettle drum, and two pairs of little kettle drums. Of the booty reserved for Prince Eugene was only the tent of the Grand Vizier, which was completely new and the most magnificent ever seen. All of the rest was abandoned to the soldiers. However, to avoid disorder and tumult, Eugene ordered that the pillaging be done by detachment in turns and with all possible order, nominating for this operation a number of sergeants and subalterns who had an eye for the soldier.

During the period of the action, the garrison of Belgrade attempted nothing. Mr. de Viard held them in respect with his countenance and by the judicious manner with which he deployed his troops.

---

[17] Translator: The *Nouveau dictionnaire historique*, Vol. 1, pp. 517-518, states that the Turks lost 13,000 dead, 5,000 wounded, and as many prisoners on the battlefield. It says the Imperials lost 3,000. The booty consisted of 130 cannon, 52 flags, nine horse tails, and four trumpets.

[18] Dumont, *Histoire militaire du Prince Eugene, Bataille de Belgrade*, p. 130.

The camels became, after the pillaging of the Turkish camp, very cheap for the Imperials, three selling for two florins. Rugs from Persia and the Indies, and the most beautiful porcelain in the world was sold for minimum prices. It would have been quite different if a few days before the battle the Turks had sent part of their baggage to Semendria.

The battle was over between 10:00 and 11:00 a.m. Prince Eugene employed the rest of the day in caring for the wounded. The next day he summoned the Pasha of Belgrade, threatening him with not quarter if he did not surrender immediately.

The Turkish officer, seeing that the outside his fortress on the side away from the river was still in a good state; that all were mined and that he could hold out for a long time, was of the opinion that he could hold out for a long time. However, the garrison, most of which consisted of married soldiers and who had witnessed the rout of the Grand Vizier, saw that there was no hope of relief and threatened to murder the Pasha if the fortress was not surrendered, believing that a longer defense was futile. On this the Pasha assembled his Divan or Council of War, at the end of which two of the garrison's officers were sent to Prince Eugene to announce to him the surrender of the city, anticipating that the garrison would be given good conditions. Eugene, who knew how much his army had suffered and how it was not in a state to continue the effort and rigors of a prolonged siege, granted everything that they wished; and the Turks, believing themselves lucky to come out without being enslaved, did not abuse his mercy. They thought only of assuring their retreat, that of their families, and their goods. The Prince granted them, in addition, the honors of war, but they did not profit from it, be it by ignorance or misunderstanding, and they came out pell-mell. The men came out overland with about 300 wagons and 1,000 camels. The women and children went by water.

The garrison was escorted to the Nissa heights by Count Philippi, Major of the Savoy Dragoon Regiment, with 50 masters.[19]

Here are the articles of capitulation, which were accorded on this day to the garrison and inhabitants of Belgrade, by His Serene Highness, Prince Eugene of Savoy:

---

[19] Translator: A "master" was an armored horseman.

I.   During the capitulation and its completion, all hostilities will cease and if any disorder arises, satisfaction will be given on both parts.

In the first place: *This is understood and one is not accustomed to act against that which is agreed.*

II.   The garrison shall engage itself to faithfully deliver the fortress in the state in which it finds itself presently, with all of the artillery, to include cannons, mortars, lead, powder, and shot, as well as all other ammunition, food, and war materials.

It is a well-known thing that everything that belongs to the defeated shall be surrendered to the victor; and that everything shall be faithfully revealed and delivered, along with the mines and ammunition.

III.   In exchange, the garrison shall be allowed to retire freely and in security with its wives, children, arms, and baggage, drums beating, and flags deployed; this shall also extend to the inhabitants, who wish to come out at the same time, of whatever condition, religion, or nation whatsoever; the same with their old slaves, which have actually embraced the Muslim faith before the siege.

Without any dispute, understanding that all of the slaves, who had been taken since the beginning of the present war, shall be surrendered without distinction; in addition, all prisoners who found themselves in the fortress taken during the siege, as well as all deserters.

IV.   And in addition, that a large part of the garrison may leave by water and that His Serene Highness, Prince Eugene, may have difficulty granting them frigates, saiques, and other ships that may serve them, His Highness is begged to grant at least the ships that cannot serve in military operations and which are not capable of anything other than facilitating the retreat, and to add there some German transports to supplement this lack; but in case that His Highness does not wish to provide such ships, because of the cannon with which they are equipped, it is offered that they be removed, such that the withdrawal not be retarded; and for this same reason one requests the assistance of some sailors.

That part of the garrison which leaves by water shall furnish itself with what ships that are in the city, and no others which can carry armament in any manner; or one will give them to the Imperial Army under the caution, as much as presently they can; and if one cannot move them all at once, the garrison shall have the liberty to assemble them in the vicinity of Varos or on an island, and leave with them some men to take care of them, as well as one who will give from our side the things necessary for the security of the garrison; thus that everything that belongs to the armament of the ships must absolutely remain and be delivered to the Imperials; one will not presently provide any sailors, and the garrison has its own saiquistes, and other men, which can handle this task; for that which is necessary for the withdrawal of the ships, the garrison will take them to Vipalanka, and will deliver them there to our territory, but if in awaiting for us to provide masters from Orsova, one shall surrender them to ours.

V.   The garrison, which shall depart by water, requests to be securely escorted by Orsova to Fetislau, beyond the narrow passage of the Iron Gates; this is because one shall leave on one side two hostages until said garrison has received the ordinary attestation that the convoy has been delivered.

This is without dispute.

VI.   As the other part of the garrison shall leave by land, one requests that they be escorted with a sufficient convoy as far as Nissa, that they be free, while on the road, both by water and land, to purchase the necessary food or sell what they will, as at Temesvar, and since they must also transport herds by land, one has need of 1,000 wagons for which and for the convoy one will leave the necessary hostages.

Although the transport by water is the most convenient, one consents, all the same, that part of the garrison shall travel by land and that it shall have the liberty to purchase food along the road and elsewhere, and to sell their effects. The provision of 1,000 wagons is completely impossible, especially considering Article IV, but nevertheless, one grants the garrison 300 wagons.

VII.   All of the prisoners who were made before and during the siege shall be surrendered and in exchange on asks that an equal response be made with regards to prisoners from the garrison.

We respond by the third article, the army has some garrisons from the garrison, on which one will make, even so, as much as one can.

VIII.   His Serene Highness shall grant to the garrison, over land, eight marches to Nissa; during which that by water shall hasten towards Fretislau as much as shall be possible, and that wind permitting, with instant prayers, that the necessary order shall be given en route, by water and land, in order that no damage or harm shall be done, by any nation, of whatever name, under whatever old pretext whatsoever.

The garrison and that which belongs to it, shall follow that which is said above, shall be escorted to a point beyond the Morave, or further if it is necessary, and then to Nissa; one shall give by the Imperial Army to that [part of the garrison moving] by ground as well as by water all security required, and one promises them, as does the escort, to take up arms against the Bleu parties that they might encounter in the field, for which purpose one shall give the necessary orders to the Imperial commanders by express couriers, and to the escort.

IX.   The sortie of said garrison shall occur in eight days, or sooner, if it can be done, and according to mutual agreement, and the exchange of the present articles, one shall initially evacuate the gate near the mosque, to the Germans; and all of the mines and munitions shall be faithfully uncovered and shown.

The sortie shall occur on the 22nd of this month, without exception, such that the capitulation shall be accepted and signed today or rejected. We do not wish to take up time with negotiations. We request, also, that after the capitulation is signed and the exchange is completed, that one gate near the mosque be evacuated with the exterior works to the right and left. One consents and permits, in addition, to the families, that find themselves in the ditches, that they shall remain in security until the actual retreat. And then one shall organize on both sides, the means to avoid all disorder. And for the security of the escorts and ships to move by water and land, one shall leave hostages who shall be returned upon the execution [of this capitulation].

Signed

Eugene of Savoy

By the order of His Serene Highness the Prince.
Von Brokhausen

Done in the Imperial camp before Belgrade,
18 August 1717.

Prince Eugene then named a Sergeant General of Battle, Count d'Olivier the interim Governor of Belgrade. The Emperor confirmed this choice.

His Serene Highness dispatched Count von Stirnm, his Adjutant General, to carry to Vienna the news of this conquest, which caused universal joy. The Imperials worked to repair the breaches and to clean the streets of debris, under which the soldiers found great riches.

A medal was struck to memorize Prince Eugene's victory.

Battle of Belgrade Commemorative Medal.

After the loss of the battle, the Grand Vizier retired on Nitza, where he could barely reassemble 30,000 men from the debris of his army.

The Turks abandoned Ram, Semendria on the Danube, and Meadia, upon the approach of Baron von Petrasch. They also abandoned Sabatz on the Save and several other places.

They gave the appearance of wishing to defend Orsova, which was a very good post, situated in mountains that were very difficult to access, but General Mercy, who Prince Eugene had sent there, dislodged them, quickly, after all of the Island of Borrez was evacuated.

Baron Petrasch then sought to capture Zwornick, in Bosnia. He initially carried the stockade by assault; but the fortress presented such a strong resistance, that he was obliged to abandon the effort.

The number of cannons and mortars captured from the Turks, be it in the city of Belgrade, in the saiques, or on the island in the Danube, and in other places was incredible. However, here is an exact accounting: In the city and castle there were 175 bronze cannon, 25 iron cannon, and 50 mortars. On the frigates and saiques there were 120 bronze cannons, 84 iron cannons, and one mortar. On the Danube island, there were 20 bronze and six iron cannon. On the saiques at that island there were 27 bronze and 27 iron cannon. In the old castle, there was one bronze and one iron cannon and eight bronze mortars.

The new of such success caused great joy in the Imperial Court. This same news produced grief and sadness in the Ottoman Court. The Grand Seigneur, whose affairs had gone no better in the Peloponnese, soon saw that it was no longer time to think of war, at least if one did not wish to completely lose it; and as this was not his intention, he resolved to buy a peace, no matter what the price. To this end he accepted the mediation of the King of England and the States-General, which Mylord Wortley Montague and Baron Colliers, their ambassadors, offered him.

However, Prince Eugene, after having had new fortifications added to Belgrade, to defend it on the water side, and after having assigned good quarters to the Imperial troops, left at the beginning of October to return to Vienna.

It is impossible to express the transports of joy that the people of this capital demonstrated upon the arrival of His Serene Highness. The Court expressed to him no less satisfaction. The Emperor wished to receive him publically, and after Eugene had given to the Emperor his show of respect, His Imperial Majesty addressed him with these remarkable words, which closed the voices of those envious of Eugene, and attempted to blacken his conduct. "The glory that you have acquired," he said, "gives you new highlights, and surpasses much that which you have acquired in your other campaigns. I thank you particularly and I shall seek occasions to give you marks of my sincere and just recognition."

Prince Eugene had barely returned to Vienna when he received a letter from the Grand Vizier, in which this prime minister assured him of the good dispositions of the Sultan, his master, for peace. Eugene communicated this letter to the Emperor and responded to the Vizier in more or less the following words: "That His Imperial and Catholic Majesty consents to establish good relations between Himself and the Sultan; but that he declares at the same time that He will not accept any treaty, except in concert with the Republic of Venice, for the defense and in favor of which, He has solely taken up arms. That this is the constant and unshakable resolution of His Imperial Majesty."

The Grand Vizier wrote anew to Prince Eugene, to inform him that the Sultan had consented to treat with the Venetians and that he was ready to send his plenipotentiaries wherever the Emperor pleased, to whom he left the liberty to name such place in Hungary where he wished to hold peace negotiations.

This letter, so different from the proud and haughty style that the Turks had always affected, caused the Viennese Court to sense the sincerity in which the Turks found themselves and moved them to propose the preliminaries which appeared exorbitant. It was Prince Eugene who proposed them in his letter to the Grand Vizier, on 15 February 1718. He said that, "as the Emperor my master claims that one establish as the basis of the negotiations the *Utipossidetis*, that is to say, that he wishes that one assure the complete and full possession of all that his arms had conquered in these two last wars. That he asks, in addition, for compensation for a war which he was forced to undertake to support his Allies, and to guarantee the frontiers of Christianity, he would abandon all of Bosnia and Serbia on the right bank of the Danube, Wallachia on the left from the Moldau River to the Dniester River."

The Sultan went into an inexpressible rage that such outrageous requests had been sent. He protested that he would sooner lose his crown than consent to a peace so withering to his reign to the most distant posterity; adding that he would sooner march all of his forces into Hungary and that one would not take it from him, even if it meant plunging so many innocent people into new misfortunes.

These threats did not shake the Viennese Court. The Emperor's council knew well that the Turks were discouraged by the outcome of the war and he judged that the Sultan's anger was the last sparks of a fire what was about to die. Nonetheless, to show that he was still in a state to continue the war, he took, on his part, to begin the same preparations for its continuation as if there had been no peace proposals.

End of the 1717 Campaign.

The Battle of Belgrade by *Jan van Huchtenburgh*

THE 1718 CAMPAIGN

It appeared that the war was about to continue with more fury than ever between the Emperor and the Porte. These two powers expressed an equal stubbornness, and the dispositions that they made were to deliver more powerful blows surpassing all that had preceded them. However, never did the Porte nor the Emperor have more need for peace. The Porte, to replace its losses; and the Emperor to drive back another power that appeared to be threatening his possessions in Italy.

The Emperor's fears were on the side of Spain, and this caused him little-by-little to surrender the articles that he had proposed to the Divan, and the Port, which only wished for peace, accepted a congress.

Passarowitz was chosen as the site for the conferences. It was a little city in Serbia, situated on the Morava.

The Emperor sent Count von Wirmont and Baron von Dalman there as his plenipotentiaries. Chevalier Ruzzini went to represent The Republic of Venice. Agas Ibrahim and Mechmet were sent as the plenipotentiaries of the Porte. Sir Robert Sutton and Baron Colliers represented the mediation of England and Holland. Even though peace negotiations had begun, the Emperor forgot nothing to rebuild his forces and to keep them capable of intimidating the Turks. He made requests to the Diet to obtain new assistance. The King of Sweden, having been chased from Pomerania and pushed into the heart of his states, and the King of Poland, the Elector of Saxony, no longer having need of his troops in the north, 6,000 Saxons were destined to reinforce the Imperial Army.

Prince Eugene prepared to take command of the newly rebuilt army. Before his departure, the nobility of Lower Austria wished to give him a brilliant mark of the high esteem in which they held him, by offering to join his corps [his lands], which was not a vain favor, since by this the Prince acquired the right to participate in the diets or estates of this province, to express his opinions, and to govern it in his turn. Prince Emanuel of Savoy, his nephew, also had part in the same favor and many joined in the growing corps of Austrian nobility, in a solemn assembly, which was held on this subject. The princes were represented by Count Leopold Victorin von Windischgrätz, who spoke brilliantly when he thanked the nobility in the name of the two princes.

This was not the only expression of appreciation that Prince Eugene received for the services he had rendered to the Empire by the capture of Belgrade and by the victory that had preceded it. When Eugene arrived in Buda and returned to Vienna at the end of this campaign, he found there Count von Rabutin, who had given him, on behalf of the Emperor, a sword whose hilt was garnished with diamonds and estimated to be worth 80,000 florins.

Prince Eugene finally left for Hungary at the beginning of June. The generals who were to serve under him had already gone there and assembled the army in the vicinity of Semlin. It was there that Prince Eugene held a review. He then visited the new fortifications that he had ordered constructed at Belgrade and he found that his orders had been perfectly followed. And as he learned that the Turks had not become superior to undertake a siege of this fortress, he thought of the means to easily secure it. This is why he had three bridges constructed over the Danube, one vis-à-vis Belgrade, one at Kuben near Semendria, and the third at the level of Orsova. At the same time, he re-established the bridge over the Save and gave orders that two more be constructed on the Morave.

In the meantime, the Grand Vizier moved on Nissa with his army. The steps taken by this general indicated to the Imperials that he did not consider himself the stronger of the two armies. He sent an aga to Prince Eugene to propose to him a suspension of arms, during which time the Passarowitz negotiations would continue. Eugene proudly rejected this proposition and responded to the aga: "That the best road to peace is to make war. That for him he was resolved to press the conclusion of a treaty by some enterprise worthy of the arms of His Imperial Majesty."

The aga returned with this unsatisfactory response and Prince Eugene made some movements to get a bit closer to the

Grand Vizier. He learned, through his spies, that the Turkish Army contained only 8,000 troops, most of them from Asia, the most cowardly that he had ever had. The rest were a mix of Arnautes, Bosnians, and Wallachians. However, as the conferences continued at Passarowitz, a significant incident occurred that delayed progress for some time. This was the examination of the full powers of the Sultan's envoys, those of the Emperor, of England and Holland commenting that they were not signed by the Grand Vizier and refused to admit them. It required that they write Constantinople and the full powers required were soon delivered. Negotiations lasted from 5 to 21 July when a treaty was signed.

The original treaty was written in Latin but here is a translation:

PEACE TREATY
Between His Majesty the Emperor, King of the Romans, and the Porte
Concluded at Passarowitz, on 21 July 1718.

In the Name of the Holy Trinity

Whatever differences have arisen in the last two years, having broken the peace that reigned between the Very August and Very Powerful Charles VI, Emperor of the Romans, and The Serene and Very Powerful Achmed Han Emperor of the Ottomans, of Asia, and of Greece, the peace which had been cemented at Karlowitz by the very glorious predecessors of these Sovereigns, and to which one had fixed the longest term, there resulted to the misfortune of their subjects and the ruin of commerce, a bloody war which caused the desolation of provinces and the ruin of peoples. However, by the grace of Divine Providence, these two emperors being returned to salutary councils, wanting to save human blood and for the welfare of their subjects, have seriously thought of a reconciliation and to remove every sentiment of animosity.

The Serene and Very Powerful King of England and Their High Powers the United Provinces having interposed to this end their good offices have convened, to negotiate for a peace treaty and to renew the former friendship to choose a location where the two sides can send their plenipotentiaries, sufficiently instructed, to negotiate the conditions.

In consequence, His Excellency, the Count Hugue-Damien de Virmont, intimate counselor of the Council of War, etc., and His Excellency Monsieur de Talman, Counselor of said council, on the part of the Serene, Most Powerful and Very Invincible Emperor of the Romans. Their Excellencies Ibrahim Aga II, President of the Ottoman Chamber and Mechmed Aga III, President of said chamber on the part of His Serene and Very Powerful Achmed Han, Grand Sultan of the Ottoman Empire; Sir Robert Sutton, in the name of His Serene and Very Powerful King of Great Britain, and His Excellency the Count Colliers in the name of Their High Power [the Dutch] have gone at the beginning of the month of May to Passarowitz and after having reciprocally shown their full powers, and having had the support necessary, they worked so efficiently on this important work, that they have agreed, on both parts, to the following 20 articles:

I. The Provinces of Moldavia and the adjoining parts of Wallachia, in part from Poland, in part from Transylvania shall be, as in the past, distinct by the mountains that separate them; such that the ancient boundaries shall be observed on all sides and that one shall make no changes be they on this side or that; and as the parts of Wallachia, situated beyond the Aluta River, as well as the Fortress of Temesvar and its dependencies are in the power of His Majesty the Emperor of the Romans, it shall conserve ownership and sovereignty, according to the principal of peace, *Uti possidetis*; such that the eastern bank of this river shall remain under the Ottomans and the western bank under the Emperor of the Romans.

II. The Aluta (Alaut) River, which flows from Transylvania, shall serve as the confines just to the point where it enters the Danube; from there along the bank of the Danube towards Ostrova to the point where the Timock River falls into this river. It shall be the Aluta as otherwise the Morosino; this to say, that it shall be common to the subjects of the two contracting parties, both to water their cattle as to fish and other commodities of this type.

It shall be permitted for the Germans and their subjects to navigate from Transylvania, with their loaded vessels, towards the two banks of the Danube, and to the subjects of Wallachia to use small boats for fishing as long as they obstruct nothing. As to the mills constructed on boats, they shall be placed in appropriate locations where they do not block navigation, and with the consent of the respective governments which find themselves on the frontiers.

In virtue of this treaty, the Bosnians and the people of lesser station, who during the war who withdrew from Ottoman Wallachia to pass to the territories of the Emperor of the Romans, shall freely return to their part to live there and enjoy as the other citizens, their former possessions.

III.   From the location where the Timock River flows into the Danube, around ten leagues above. It is necessary to fix the limits of the two empires. In consequence Isperleckbania, with its ancient territory shall remain with the Ottomans and Ressowa with His Imperial Majesty. From there one shall advance between the mountains towards Parakin, such that this latter location shall remain under the domination of the Emperor and Rasna under that of the Porte. In following the middle of these two locations the frontiers shall pass Istolaz, the Little Morava, and follow the river going to Schahak, and enter Shahak and Bilana crossing the fields to Bedka: from there turning towards the territory of Zoklense to Bellina situated on the banks of the Drina River; such that Belgrade, Parakin, Istaloz, Schahak, Bedka, and Bellina with their ancient territories remain in the power of His Majesty the Emperor of the Romans, since he possesses them; Zokol and Rasna shall continue to recognize the Ottoman Empire. The subjects of the two empires shall enjoy the commodities that are officered by the Timock River.

IV.   From the vicinity where the Unna falls into the Sava to the territory of Antiqui-Novi situated on the eastern bank of this river and belonging to the Ottoman Porte, Jassenowitz, Dobiza, the towers and some islands being occupied by the troops of the Emperor of the Romans shall remain in the power of His Majesty according to the principal of this peace treaty.

V.   The territories of Novi-Novi situated on the western bank of the Unna on the side of Croatia shall otherwise belong to the Emperor of the Romans, having been, according to the Treaty of Karlowitz, as the result of various different issues that arose, given to the Ottoman Empire, after having ruined the stockades; they shall be in the form of reconciliation and satisfaction, returned to His Majesty the Emperor of the Romans, as well as all of the territories that find themselves between the ancient frontiers.

VI.   In addition, all of the territories of Croatia along the Save that each of the two powers possess and where they have placed garrisons, shall remain, with their dependencies, as established in the Treaty of Karlowitz, in the power of the occupant who conquered them, but in 24 years (lunar) to count from the day of the ratification of the present treaty, the two contracting parties shall send deputies to fix the limits and cut the root of all dissension; they shall go to the extremity of

Croatia; they shall grant on these territories which shall remain to each of the two powers, fixing them in a precise and clear manner and by unequivocal limits. It is permitted by the present treaty, as by that of Karlowitz, to each of the two powers to repair and fortify for its own security the castles and fortresses which they possess; the same to the inhabitants of the confines of the two empires, without exception, to construct in the open regions convenient habitations, such that they not serve as a pretext to construct new fortresses.

VII.   By the means of the conditions stated above in the present peace treaty were concluded in good harmony, nonetheless so that what was promised, and accepted compared to borders shall acquire greater forces the two contracting parties must each send expert commissioners, faithful, and peaceful who, followed by such servants as they will need, will meet in a place that they judge most appropriate (in the space of two months and sooner if possible) to place distinctive marks on the frontiers designated in the preceding articles; and all of that which shall be resolved by the two sides shall be followed with the most exact and prompt execution.

VIII.   The limits fixed by the present treaty and to be posted by the commissioners of the two contracting powers, where it shall be needed, as it has been stated above, or those that said commissioners determine as a result, and in a most opportune time, shall religiously and faithfully be observed by both sides, such that these limits shall not be changed, nor extended in any manner whatever or under whatever pretext might be.

In addition, after these limits are fixed, there shall not be permitted that either party shall claim any right, nor exercise any violence on the territory of the other, nor demand of the subjects of the other party any tribute, nor tax, be it for the past, the present, or the future, nor exercise any type of exaction or vexation whatever but on the contrary one will carefully eliminate all subjects of dispute.

IX.   To eliminate all difficulties that might arise on the borders that are subject to any article of this peace, and to appease any discord that might arise, one shall name from the two parties, when circumstances demand, some commissioners, grave, men of probity and prudence, experts and peaceful, and in an equal number, who shall assemble in an appropriate location, to this end, without troops, and with an equal number of peaceful men; they shall know all difficulties of this type and arrange them in an amicable manner after having made all necessary research, and they shall establish order in such a manner that each party shall oblige their subjects, in whatever case may be, to a rigid observation of the peace.

However, if there arrives a case of such great importance that the commissioners of the two parties cannot come to a decision then the situation shall be passed to the two Very Powerful Emperors, in order to decide on the means necessary to remedy the situation, to smooth it over, and to eliminate all difficulties, such that these discords are settled as quickly as possible, and one cannot neglect them for any reason.

As duels and cartels have been prohibited in the previous capitulations, so shall they be prohibited in the future, and those who commit transgressions shall be very severely punished.

X.   All hostilities, incursions, and forceful attacks either open or by surprise, the devastation and the depopulation of territory of the two powers are severely prohibited; and those who make themselves culpable shall be taken anywhere one finds them, imprisoned, and judged by the jurisdiction where they shall be arrested, and shall not be susceptible to any mercy. As to robbery, of whatever nature, one shall rigorously search [for the robbers] and if they are found, there shall be restitution made to the proprietor. The commandants, governors, and custom's officers of the two powers shall hold the administration of justice with the greatest integrity, not only under the pain of losing their employment, but also their lives and honor.

XI. The Serene Emperor of the Ottomans shall confirm all that his glorious predecessors have done in favor of the religious and the Roman Catholic Religion, be it by previous capitulations, or by edicts or particular orders; of such that said religious people shall be able to repair and rebuild their churches and exercise the functions of their ministry as in the past. There shall not be permitted that anyone molest them, against the will of the law, and against older capitulations, nor to demand any retribution against those religious, of whatever order they may be; but to the country they shall enjoy as before the protection of the Serene Emperor of the Ottomans. In addition, it shall be permitted to the envoy of the Very Powerful Emperor of the Romans to the Ottoman Porte to present his commissions, relative to religion, in the Holy Land, and to all of the places where said religious may have churches, and to make entreaties on this subject.

XII. The prisoners made during the preceding war and in it, by either party, which are still today deprived of their liberty, as a result of this peace treaty, shall recover their liberty; they cannot, without attacking Imperial clemency, and derogating from creditable use and generosity, remain any longer in a state of captivity or suffering. As a result, all of the reciprocal prisoners, as in the past, shall be placed at liberty in 61 days. However, one shall exchange immediately Nicolas Scarlati Voivode, his son, and his servants detained as prisoners in Transylvania against Barons Petrasch and Stein, and their people imprisoned in the seven towers in Constantinople, and they shall be, in the space of 31 days, taken to the borders of Wallachia to be handed over to the power of their respective sovereigns.

As for other prisoners who are in the power of individuals or Tartars, he shall permit them to seek to purchase their freedom at an honorable price, and a low price if possible. If one cannot come to a reasonable price with the masters of the captives, it is to the judges in these cases to grant them privately. However, if this way is not practical, the captives can by witness or oath of the price, by which they were bought and reimburse it, shall be set free and the avidity of their masters shall not oppose it.

That if some part of the Ottoman Empire has not sent men charged with facilitating the ransom of captives it shall be at the integrity of the judges of the Emperor of the Romans to oblige those who have prisoners to surrender them for the price that they paid for them, in order that one contribute equally from the two sides to a good work.

And up to the moment when the unfortunate prisoners are respectively placed at liberty, the plenipotentiaries of the two powers shall direct their cares to assure that they are treated kindly.

XIII.   According to preceding treaties the merchants of the two powers may safely exercise commerce in the two empires. The subjects of His Majesty the Emperor of the Romans and those of the provinces of the Christian states which as a result may pass under his domination shall issue passports by sea and by land to go and to return as they have been regulated by the commissioners provided that the ships carry the Imperial flag or that the merchants be equipped with letters of patent from His Majesty the Emperor and then he shall permit them to buy and sell in the different provinces of the Ottoman Empire and in the kingdoms that are under its domination; and far from one molesting them, having paid all of the established taxes, they shall enjoy complete protection. One shall establish in convenient places between the Consul Commissioners and the interpreters charged with assuring the well-being of commerce, and the merchants of the states of His Majesty the Emperor of the Romans shall enjoy complete security, of all favors, and of all the advantages accorded to all other Christian nations that are exempted from tribute.

One will direct Algiers, Tunis, Tripoli, and other people who practice piracy to undertake nothing in opposition to the conditions of this peace treaty. One shall prohibit, in addition, the Dulcinotes from attacking in the future merchant ships; or taking their frigates and preventing all construction, and those who in the future attempt to contravene this treaty and attack merchant ships of the Emperor of the Romans, shall be held to an indemnity, to the restitution of captives, and their punishment shall be exemplary.

XIV.   It shall always be prohibited to give asylum to rebels or discontents, and the two powers shall oblige themselves to have them executed, as well as all thieves or ravishers, taken in their territories, even if they are subjects of the other party. If they escape and one knows of their retreat or denounces them to the judges who shall hold them to be punished; and if they are found guilty of negligence in this regard, they shall be disgraced by their sovereign, and they shall be deprived of their employment or punished in the same place as the criminals.

And to cut the root of all excess neither of the two parties shall give asylum or subsistence to the Haydons, to the Pribecks, nor any other people with a sovereign, nor to any who are not in the pay of one or the other of the two powers, and who live by theft. Those who give them assistance shall be punished according to the demands of the case. If some of these vagabonds give hope of rehabilitation, they shall not be suffered on the frontiers, but transported to distant locations.

XV.   So that the borders not be exposed to be troubled in any manner, Ragotzi, Berezeni, Antoine Esterhazy, Forgatsch, Adam Vai, and Michel Zachy, as well as all Hungarians who in the time of war rejected obedience to the Emperor of the Romans, and have sought asylum in the territories of the Ottomans, they shall be transported to the locations where the Porte may please to assign them, but this cannot be in the vicinity of the frontiers. They shall be permitted to have their wives follow them and live with them.

When the plenipotentiaries of His Majesty the Emperor have proposed that the King and the Republic of Poland be included in this treaty, he responded that there is a perpetual peace with the King and the Republic, and that they have no differences with the Porte. That if, however, the Poles have some proposition to make relative to Chocim, or otherwise, they can treat by ambassadors in writing with the Porte which will respond with all justice.

XVII. To confirm all the better with this peace and the amity between the two Powerful Emperors, the ambassadors who shall be reciprocally sent with the ordinarily ceremonies, shall receive with the accustomed honors and they carry voluntary presents, nonetheless conforming to the dignity of the two powers. They shall set out on the spring equinox and after a mutual correspondence, they shall arrive at the same time on the frontiers and exchange according to the ancient usages of the two empires. It shall be permitted to the two ambassadors to make requests which may please them in the respective courts where they are sent.

XVIII. As to the honors to render to the two ambassadors during their voyage and sojourn, one will follow the old form; and the ceremony shall be equal on the two sides. But one shall observe the character and the rank of the envoys sent to conform to it. There shall be permitted to the ministers, residents of His Majesty the Emperor of the Romans, as well as to all in their suite to follow the custom that they wish, without any opposition. In addition, said ministers, be they ambassadors, residents or chargés d'affaires, shall enjoy the same privileges as all other ministers of courts friendly to the Porte, and the same with whatever distinctions with regard to the Imperial dignity. They shall have, in addition, permission, to maintain interpreters. Their couriers and those that they may expedite to their court or to others shall enjoy on their route all security possible, and one will procure for them the same security that they may require.

XIX. The plenipotentiaries of the two powers shall oblige to have the present articles ratified by their respective masters in the space of 31 days, and sooner, if possible, to count from the day where they are signed, and also from the mutual exchange.

XX. This peace shall, by the Grace of God, last for 24 years; then and even before the term, these two great contracting powers may prolong it.

All of the above that has been voluntarily ruled by His Majesty the Emperor of the Romans and His Majesty the Emperor of the Ottomans shall be faithfully observed by their heirs, empires, kingdoms, lands, maritime shores, republics, cities, and subjects who recognize their domination.

One shall seriously inform the governments of the two parties, their troops, and all those who are under their power, that they must conform to all of the aforementioned articles, and that they have to take care not to do anything which might breach this peace and with the friendship that reigns between the two Great Powers, but on the contrary in laying down any kind of enmity they shall live as good neighbors, persuaded that if they contravene any of these articles they will be very severely punished.

The Khan of Crimea and all of the Tartar nations, without exception, shall observe all of the conditions of this treaty and live in good harmony with all of the subjects of His Majesty the Emperor of the Romans and if anyone, be it a soldier or Tartar, dare to contravene these imperial capitulations and do something contrary to this pact and its articles, they shall be rigorously punished.

Said peace and with it the repose and the security of the subjects shall begin on the day of the signing of the present pact and to that end all hostilities shall cease between the subjects of the two empires. And to reach this as quickly as possible, one shall immediately send and with the greatest celerity orders to all of the governors of the borders in their announcement of this peace; but as it will require some time to inform the officers who are in the most distant provinces, one shall fix the term of 20 days after which all hostile action, be it on one part or the other, will incur the sorrows above stated and will not be given any forgiveness.

And so that the 20 articles of the concluded peace and accepted by the two powers are completely observed, after which the plenipotentiaries of the Ottoman Porte have shown us their full power written in the Turkish language and signed, we have, with the authority that has been given to us, drawn up the present treaty in the Latin language, and have placed on it the seal with which we are armed and have signed with our own hand as the legitimate and valid instrument.

Done in the congress assembled at Passarowitz, under the tents erected to this end, the 21st of July 1718.

S. Damien Hugue
Count de Virmont

S. Michel de Talman

Once the treaty was signed a courier was sent to Prince Eugene to inform him of its signing. His Highness was then on the march, looking to find the Grand Vizier and bring him to battle. This rubbed him the wrong way, despite the great hopes he had for attacking the Sultan in his capital and obliging the Muslims to surrender to the Christians an empire which they only held because of its divisions. He was surprised that a cardinal from one of the first columns of the Church, could abuse the confidence of his master to make war on the Emperor while this monarch was on the verge of destroying the Sultan.

The pretext taken by Spain to attack Charles VI was to prevent the exchange which this prince wished to make with the Duke of Savoy, of Sardinia for Sicily, the possession of which had been assured by the Treaty of Baden.

However, Prince Eugene went to occupy himself to arrange districts for the Imperial troops and to strengthen the most exposed fortresses. By his cares, Orsova was to become almost impregnable and Belgrade, because of the profound knowledge that Eugene had of the art of fortification, became regarded as an unassailable fortress.

Finally, Prince Eugene having nothing more to do in Hungary, returned to Vienna, where his councils were of great use considering the state of affairs.

Always ready to render the greatest services to the House of Austria, he was no less useful in the Cabinet as at the head of its armies. Eugene, with all of the qualities of a hero, had all of the virtues of an individual. Modest in the heart of victory, seeking quiet, even in the tumult of a capitol, gentle, human, and kindly, he had the charms of friendliness and enjoyed his last years as a great man. He died at age 72 years, 6 months, and 18 days at a time when his wisdom and courage would be greatly needed by Austria, that is to say, a year before the rupture of the Treaty of Passarowitz, which was the result of his victories. Tears of appreciation frequently flew on the mausoleum which was raised for him by the Emperor and his death caused general grief.

Frederick Wilhelm Karl von Schmettau

Select Biographies

Frederick Wilhelm Karl von Schmettau

Eugene of Savoy by *Johann Gottlieb Auerbach*

Eugene of Savoy (1663 –1736) better known as Prince Eugene, was a field marshal in the army of the Holy Roman Empire and of the Austrian Habsburg dynasty during the 17th and 18th centuries. He was one of the most successful military commanders of his time and rose to the highest offices of state at the Imperial court in Vienna. He was brought up in the court of King Louis XIV of France. Although initially destined for a clerical career, but by the age of 19, he had determined on a military career. He was rejected by Louis XIV for service in the French army and transferred his loyalty to the Holy Roman Empire.

His first battle experiences were fought against the Ottomans at the Siege of Vienna in 1683 and the subsequent War of the Holy League, before serving in the Nine Years' War, in which he fought alongside his cousin, the Duke of Savoy. Eugene won fame in his victory against the Ottomans at the Battle of Zenta in 1697. During the War of the Spanish Succession, his partnership with the Duke of Marlborough secured victories against the French at of Blenheim (1704), Oudenarde (1708), and Malplaquet (1709). He gained further success in the war as Imperial commander in northern Italy, most notably at the Battle of Turin (1706). Renewed hostilities against the Ottomans in the Austro-Turkish War consolidated his reputation, with victories at the battles of Petrovaradin (1716), and the decisive encounter at the Siege of Belgrade in 1717.

Throughout the late 1720s, Eugene's influence and skillful diplomacy managed to secure the Emperor powerful allies in his dynastic struggles with the Bourbon powers, but physically and mentally fragile in his later years, Eugene enjoyed less success as commander-in-chief of the army during his final conflict, the War of the Polish Succession. Nevertheless, in Austria, Eugene's reputation remains unrivalled. Although opinions differ as to his character, there is no dispute over his great achievements: he helped to save the Habsburg Empire from French conquest; he broke the westward thrust of the Ottomans, liberating parts of Europe after a century and a half of Turkish occupation; and he was one of the great patrons of the arts whose building legacy can still be seen in Vienna today. Eugene died in his sleep at his home on 21 April 1736, aged 72.

Charles VI by *Johan Gottlieb Auerbach*

Charles VI of Austria (1685 – 1740) was Holy Roman Emperor and ruler of the Austrian Habsburg monarchy from 1711 until his death, succeeding his elder brother, Joseph I. He unsuccessfully claimed the throne of Spain following the death of his relative, Charles II. In 1708, he married Elisabeth Christine of Brunswick-Wolfenbüttel, by whom he had his four children: Leopold Johann (who died in infancy), Maria Theresa (the last direct Habsburg sovereign), Maria Anna (Governess of the Austrian Netherlands), and Maria Amalia (who also died in infancy).

Four years before the birth of Maria Theresa, faced with his lack of male heirs, Charles provided for a male-line succession failure with the Pragmatic Sanction of 1713. The Emperor favored his own daughters over those of his elder brother and predecessor, Joseph I, in the succession, ignoring the decree he had signed during the reign of his father, Leopold I. Charles sought the other European powers' approval. They demanded significant terms, among which were that Austria close the Ostend Company. In total, Great Britain, France, Saxony-Poland, the Dutch Republic, Spain, Venice, The Papal States, Prussia, Russia, Denmark, Savoy-Sardinia, Bavaria, and the Diet of the Holy Roman Empire recognized the sanction. France, Spain, Saxony-Poland, Bavaria and Prussia later reneged. Charles died in 1740, sparking the War of the Austrian Succession, which plagued his successor, Maria Theresa, for eight years.

Ahmet III

Ahmet III (1673 –1736) was Sultan of the Ottoman Empire and a son of Sultan Mehmed IV (r. 1648–1687). His mother was Gülnuş Sultan, originally named Evmenia Voria, who was an ethnic Greek. He was born at Hacıoğlu Pazarcık, in Dobruja. He succeeded to the throne in 1703 on the abdication of his brother Mustafa II (1695–1703). Nevşehirli Damat İbrahim Pasha and the Sultan's daughter, Fatma Sultan (wife of the former) directed the government from 1718 to 1730, a period referred to as the Tulip Era.

The first days of Ahmed III's reign passed with efforts to appease the janissaries who were completely disciplined. However, he was not effective against the janissaries who made him sultan. Çorlulu Ali Pasha, who Ahmed brought to the Grand Vizier, tried to help him in administrative matters, made new arrangements for the treasury and Sultan. He supported Ahmed in his fight with his rivals.

Frederick Wilhelm Karl von Schmettau

INDEX

Look for more books from Winged Hussar Publishing, LLC – E-books, paperbacks and Limited-Edition hardcovers. The best in history, science fiction and fantasy at:
https://www. wingedhussarpublishing.com
https://www.whpsupplyroom.com
or follow us on Facebook at:
Winged Hussar Publishing LLC
Or on twitter at:
WingHusPubLLC
For information and upcoming publications

Maurice de Saxe's 1745 Campaign in Belgium

By Henry Pichat
Translated by George Nafziger

THE RUSSIANS AND PRUSSIANS DURING THE SEVEN YEARS WAR
BY
ALFRED RAMBAUD

TRANSLATED BY
GEORGE NAFZIGER

The Legion Du Nord 1806 - 1808
By
Major Lazare Claude Coqueugniot

Translated by
George Nafziger

www.ingramcontent.com/pod-product-compliance
Lightning Source LLC
Chambersburg PA
CBHW041012140426

R18136400002B/R181364PG42813CBX00011B/3